Scientific Basis of
Herbal Medicine

The Editor

Dr. Parimelazhagan Thangaraj, is Associate Professor in the Department of Botany, School of Life Sciences, Bharathiar University, Coimbatore. His areas of specialization include Bioprospecting of medicinal plants. He has worked as Scientist in DRDO, Govt. of India. He has filed four patents and has more than 35 research publications to his credit. He is working on natural products for the past 10 years and guiding 8 Ph.D. scholars. His book 'Herbal Perspective: Present and Future' is an excellent reference book for those biologists and pharmacy students. He has four books to his credit on various aspects of medicinal plants and filed 4 patents. He has bagged National Science Day Medal 2005 from DRDO HQ, New Delhi and Laboratory Scientist of the Year award 2005 from DRDO.

Scientific Basis of
Herbal Medicine

Editor

Dr. Parimelazhagan Thangaraj

Associate Professor
Department of Botany,
School of Life Sciences
Bharathiar University,
Coimbatore, T.N., India

2013
Daya Publishing House®
A Division of
Astral International Pvt. Ltd.
New Delhi – 110 002

Published by	:	**Daya Publishing House®**
		A Division of
		Astral International Pvt. Ltd.
		– ISO 9001:2008 Certified Company –
		4760-61/23, Ansari Road, Darya Ganj
		New Delhi-110 002
		Ph. 011-43549197, 23278134
		E-mail: info@astralint.com
		Website: www.astralint.com
Laser Typesetting	:	**Classic Computer Services**
		Delhi - 110 035
Printed at	:	**Salasar Imaging Systems**
		Delhi - 110 035

PRINTED IN INDIA

Dr. G. James Pitchai
Vice-Chancellor

BHARATHIAR UNIVERSITY
State University
Coimbatore – 641 046

Foreword

South India FOREWORD Traditional medicinal knowledge of medicinal plants and their use by indigenous cultures are not useful for conservation of cultural traditions and biodiversity but also for community healthcare and drug development in the present and future. Plant-based traditional knowledge has become a recognized tool in search for new sources of drug and neutraceuticals and the ethnobotanical survey can bring out many different clues for the development of drugs to treat human diseases. I am happy that an exclusive book, which exemplifies the potential and recent trends on various aspects of herbal medicine, through manuscripts contributed by eminent researchers has been attempted. This book provides a ground work for natural product chemistry/phytochemistry, pharmacology, natural antioxidant and standardization of herbal medicines leading to expansion of knowledge in various selected classes of herbal science. I congratulate Dr. T. Parimelazhagan for his tremendous efforts in bringing out this inclusive publication on *"Scientific Basis of Herbal Medicine"*. It is intently believed that students, learning herbal science will certainly find this text not only useful but also a good companion for further pursuit of higher knowledge. I am sure this book will be welcomed and provide scientific guidelines for the researchers, academicians and students as a reference work for all aspects of herbal research in India.

G. James Pitchai

Ph (O): 091-422-2422439, (R): 091-422-2422500, Fax: 091-422-2422387
E-mail: vc@b-u.ac.in, website: www.b-u.ac.in

Preface

Medicinal plants have acquired increasing significance in recent years. Their use and conservation are cross-sectoral concerns that embrace not only health care but also nature conservation, biodiversity, economic assistance, trade and legal aspects (*e.g.* intellectual property). The increasing popularity in plant-based drugs is now felt all over the world leading to a fast growing market for plant based drugs, pharmaceuticals, nutraceuticals and functional foods. The principal aim of this book has been to collate the latest information and, review the viewpoints of eminent investigators, who excel in the areas of herbal drug research. An attempt has been made to compile most relevant and recent trends of scientific information pertinent to ethnobotany, ethnopharmacology, phytochemistry, bioinformatics and biotechnology. This book is a compilation of 28 articles by eminent academicians and scientists on different aspect of the subject. All the recent scientific innovations/advances have been incorporated in a simple and lucid manner. A special emphasis has been laid on screening of herbal drugs for pharmacological activity, antioxidant potential, phytochemistry and bioinformatics tools involved in herbal research. The review processes of the articles have been carried out by experts from Universities and Research Institutes. We hope the present compilation will be useful for the Students, Research Scholars, Academicians and Industrialists and people associated with herbal research. We appreciate the efforts of all those who contributed the quality papers by sharing their current scientific knowledge. The editor would like to convey sincere thanks to Dr. G. James Pitchai, Vice-Chancellor, Bharathiar University for his guidance and encouragement. The support and guidance rendered by Dr. V. Narmatha Bai, Professor & Head, Department of Botany, Bharathiar University is also acknowledged. The author would also like to thank his research team Dr. R. Senthil Kumar, Blassan P. George, M. Iniyavan, K. Arunachalam, S. Saravanan, Rahul Chandran and Sajeesh T. for their contribution in the compilation of the manuscripts in a well designed manner. The extends his sincere thanks to Department of Science and Technology, Ministry of Earth Science, Defence Research Development Organisation (Govt. of India), Tamil Nadu State Council for Science and Technology (Govt. of Tamil Nadu) for providing financial support to conduct "National Conference on Phytomedicine".

Constant support provided by our colleagues of our Department of Botany, Bharathiar University, Coimbatore is also acknowledged. We would like to express our special appreciation for the publishers and their team for the sincere efforts in bringing out the book in time.

Parimelazhagan Thangaraj

Editor

Contents

List of Contributors

Ahamed, M. Faizal
Department of Biotechnology, Achariya Arts and Science College, Villianur, Puducherry – 605 110

Anbu, Thangadurai
Research Department of Botany, Sri Parasakthi College for Women (Autonomous), Courtallam, Tamil Nadu

Aravinthan, K.M.
Research Department of Biotechnology, Dr. Mahalingam Center for Research and Development, NGM College, Pollachi – 642 001

Aron, Somu
Research Department of Botany, V.H.N. Senthikumara Nadar College (Autonomous), Virudhunagar – 626 001, Tamil Nadu, India

Girishkumar, E.
P.G. Dept. of Plant Science, Mahatma Gandhi Govt. Arts College, Mahe – 673 310, U.T. of Puducherry

Gopika, S.
Department of Botany Avinashilingam University, Coimbatore– 43

Govindarajan, M.
Division of Vector Biology and Phytochemistry, Department of Zoology, Annamalai University, Annamalai Nagar – 608 002, Tamil Nadu, India

Harilal, C.C.
University of Calicut, Calicut University P.O. – 673 635, Kerala

Hemalatha, K.
Department of Botany Avinashilingam University, Coimbatore– 43

Johnson, M.
Centre for Plant Biotechnology, Department of Plant Biology and Plant Biotechnology, St. Xavier's College (Autonomous), Palayamkottai, Tamil Nadu, India

Kamalanathan, Desingu *et al.*
Natural Drug Research Laboratory, Department of Biotechnology, Periyar University, Salem – 636 011, Tamil Nadu, India

Kathiresh, M.
Research Department of Biotechnology, Dr. Mahalingam Center for Research and Development, NGM College, Pollachi – 642 001

Kumar, B. Senthil
Department of Biotechnology, Kumaraguru College of Technology, Coimbatore – 641 049, Tamil Nadu, India

Kumudha, P.
Department of Botany Avinashilingam University Coimbatore – 641 043, Tamil Nadu, India

Lalitha, R.
Department of Biotechnology, Achariya Arts and Science College, Villianur, Puducherry – 605 110

Latha, A.
Department of Bioinformatics, Sri Krishna Arts and Science College, Coimbatore

Malar, T. Renisheya Joy Jeba
Centre for Plant Biotechnology, Department of Plant Biology and Plant Biotechnology, St. Xavier's College (Autonomous), Palayamkottai, Tamil Nadu, India

Manojkumar, A. *et al.*
Department of Biotechnology, Kumaraguru College of Technology, Coimbatore – 641 049, Tamil Nadu, India

Meenaskhi, G.J. Sree
Department of Biotechnology, Kumaraguru College of Technology, Coimbatore – 641 049, Tamil Nadu, India

Mehalingam, Palanichamy
Research Department of Botany, V.H.N. Senthikumara Nadar College (Autonomous), Virudhunagar – 626 001, Tamil Nadu, India

Mehalingam, Palanichamy
V.H.N. Senthikumara Nadar College (Autonomous), Virudhunagar, Tamil Nadu

Muthudharani, R.
Department of Biotechnology, Kumaraguru College of Technology, Coimbatore – 641 049, Tamil Nadu, India

Natarajan, Devarajan
Natural Drug Research Laboratory, Department of Biotechnology, Periyar University, Salem – 636 011, Tamil Nadu, India

Padmaja, C.K.
Department of Botany, Avinashilingam University, Coimbatore – 43, Tamil Nadu, India

Paulsamy, Subramanium
Department of Botany, Kongunadu Arts and Science College, Coimbatore – 641 029

Pradeepkumar, G.
P.G. Dept. of Plant Science, Mahatma Gandhi Govt. Arts College, Mahe – 673 310, U.T. of Puducherry

Priscilla, H. Deena
Department of Microbiology, Cauvery College for Women, Tiruchirappalli – 620 018, Tamil Nadu, India

Priya, K. Shanmuga
Department of Biotechnology, Kumaraguru College of Technology, Coimbatore – 641 049, Tamil Nadu, India

Rajeswary, M.
Division of Vector Biology and Phytochemistry, Department of Zoology, Annamalai University, Annamalai Nagar – 608 002, Tamil Nadu, India

Rakkimuthu, R.
Research Department of Biotechnology, Dr. Mahalingam Center for Research and Development, NGM College, Pollachi – 642 001

Ramachandran, R.
Department of Biotechnology, Achariya Arts and Science College, Villianur, Puducherry – 605 110

Ramalingam, P.
Department of Biotechnology, Kumaraguru College of Technology, Coimbatore – 641 049, Tamil Nadu, India

Ramalingam, P.
Department of Biotechnology, Kumaraguru College of Technology, Coimbatore – 641 049, Tamil Nadu, India

Ranjani, M.
Department of Microbiology, Cauvery College for Women, Tiruchirappalli – 620 018, Tamil Nadu, India

Rashmi, B. Arthi
Department of Bioinformatics, Sri Krishna Arts and Science College, Coimbatore

Ravindran, C.P.
P.G. Dept. of Plant Science, Mahatma Gandhi Govt. Arts College, Mahe – 673 310, U.T. of Puducherry

Reshmi, S.K.
Research Department of Biotechnology, Dr. Mahalingam Center for Research and Development, NGM College, Pollachi– 642 001

Saraswathy, N.
Department of Biotechnology, Kumaraguru College of Technology, Coimbatore – 641 049, Tamil Nadu, India

Saraswathy, N.
Department of Biotechnology, Kumaraguru College of Technology, Coimbatore – 641 049, Tamil Nadu, India

Sasikala, K.
P.G. Dept. of Plant Science, Mahatma Gandhi Govt. Arts College, Mahe – 673 310, U.T. of Puducherry

Senguttuvan, Jamuna
Department of Botany, Kongunadu Arts and Science College, Coimbatore – 641 029

Shanmugapriya, K.
Dr. G. R. Damodaran College of Science, Coimbatore – 641 014, Tamil Nadu, India

Shree, D. Ramya
Department of Biotechnology, Kumaraguru College of Technology, Coimbatore – 641 049, Tamil Nadu, India

Sindhu, S.
Department of Microbiology, Cauvery College for Women, Tiruchirappalli – 620 018, Tamil Nadu, India

Singh, Anju
Department of Botany Avinashilingam University Coimbatore – 6410 43, Tamil Nadu, India

Sivakumar, R.
Division of Vector Biology and Phytochemistry, Department of Zoology, Annamalai University, Annamalai Nagar – 608 002, Tamil Nadu, India

Sivaselvi, P.
Department of Bioinformatics, Sri Krishna Arts and Science College, Coimbatore

Stalin, N.
Department of Plant Sciences, School of Biological Sciences, Madurai Kamaraj University, Madurai– 625 021, Tamil Nadu, India

Sundaravel, K.
Department of Biotechnology, Achariya Arts and Science College, Villianur, Puducherry – 605 110

Swamy, P.S.
Department of Plant Sciences, School of Biological Sciences, Madurai Kamaraj University, Madurai– 625 021, Tamil Nadu, India

Thamilmaraiselvi, B.
Department of Microbiology, Cauvery College for Women, Tiruchirappalli – 620 018, Tamil Nadu, India

Thayumanavan, T.
Dr. G. R. Damodaran College of Science, Coimbatore – 641 014, Tamil Nadu, India

Theivandran, G.
Department of Zoology, VOC College, Millerpuram, Tuticorin – 628 008, Tamil Nadu, India

Wilson, Shiji *et al.*
Department of Botany, Avinashilingam University, Coimbatore – 43, Tamil Nadu, India

Yogalakshmi, K.
Division of Vector Biology and Phytochemistry, Department of Zoology, Annamalai University, Annamalai Nagar – 608 002, Tamil Nadu, India

Scientific Basis of Herbal Medicine (2013)
Editor: Dr. Parimelazhagan Thangaraj
Published by: DAYA PUBLISHING HOUSE, NEW DELHI

Pages 1–8

Chapter 1

Phytochemical Analysis and Antimicrobial Activity of *Alstonia scholaris* R.Br and *Rauvolfia serpentina* (L.) Benth. (Apocynaceae)

Thangadurai Anbu[1] and Palanichamy Mehalingam[2]

[1]Research Department of Botany,
Sri Parasakthi College for Women (Autonomous),
Courtallam, Tamil Nadu, India
[2]V.H.N. Senthikumara Nadar College (Autonomous),
Virudhunagar, Tamil Nadu, India

1. Introduction

The Apocynaceae family consists of about 250 genera and 2000 species of tropical trees, shrubs and vines. This family is known for plants that have a very high biological activity and medicinal properties like treatment of disorders of skin, liver diseases, leprosy, dysentery, ulcers, ear aches and fever (Holds worth, 1986). *Alstonia scholaris* R.Br and *Rauvolfia serpentine* (L.) Benth. *ex* Kurz also reported that to have anti-cancer properties, the latex being used widely to cure sores, ulcers, tumours and dropsy (Zashim *et al.*, 2006).

Plants have been an important source of medicine for thousands of years. Even today, the World Health Organization (WHO) estimated that up to 80 per cent of people depend on traditional medicine. Its civilization is very ancient and the country as a whole has long been known for its rich resources of medicinal plants. Today, Ayurvedic, Homeopathy and Unani physicians utilize numerous species of medicinal plants that found their way a long time ago into the Hindu Materia Media (Narayana and Thammanna, 1987).

In recent years, secondary plant metabolites alkaloids, anthroquinones, coumarins, phenols, flavones, tannins, saponins, steroids, terpenoids, xanthoproteins, quinines and sugars are previously with unknown pharmacological activities, have been extensively investigated as a source of medicinal agents (Krishnaraju *et al.*, 2005). Thus, it is anticipated that phytochemicals with adequate antibacterial efficacy will be used for the treatment of bacterial infections (Balandrin *et al.*, 1985). Since time immemorial, man has used various parts of plants in the treatment and prevention of various ailments (Tanaka *et al.*, 2002).

The aim of this study was to evaluate the phytochemical analysis and antimicrobial activity of methanol extracts from 2 plants against Gram-positive, Gram-negative bacteria and fungus *Aspergillus niger* strains *in vitro*.

2. Materials and Methods

2.1. Collection and Authentication of Plant Materials

Fresh plant/plant parts of *Alstonia scholaris* R.Br and *Rauvolfia serpentina* (L.) Benth. *ex* Kurz were collected randomly from Western Ghats of Courtallam in Tamil Nadu. The taxonomic identities of these plants were confirmed by Flora of the Presidency of Madras (Gamble, 1925).

2.2. Preparation of Extracts

Ten grams of the shade dried leaf powdered samples of the above plants were extracted in 40 ml of 80 per cent Methanol in a Soxhlet apparatus and used for phytochemical analysis. All the extracts were subjected to qualitative tests for the identification of various phytochemical constituents as per the standard procedures procedures (Treas and Evans, 1978; Lala, 1993).

2.3. Test for Alkaloids

The test solution was mixed a little amount of dilute hydrochloric acid and Mayer's reagent. Formation of the white precipitate indicated the presence of alkaloids.

2.4. Test for Anthroquinones (Borhtrager's Test)

To the test solution a few drops of magnesium acetate solution was added. Formation of pink colour indicated the presence of anthroquinones.

2.5. Test for Coumarins

To 2 ml of the test solution, a few drops of alcoholic sodium hydroxide was added. Formation of yellow colour indicated the presence of coumarins.

2.6. Flavones (Shindos Test)

To the test solution, a few magnesium turnings and a few drops of concentrated hydrochloric acid were added and boiled for five minutes. Formation of red or orange red colour indicated the presence of flavones.

2.7. Phenols

To the test solution, a few drops of ferric chloride solution were added. Formation of bluish green or red colour indicated the presence of phenols.

2.8. Quinones

The test solution was treated with few drops of concentrated sulphuric acid or aqueous Sodium hydroxide solution. Color formation indicates presence of quinone compounds.

2.9. Test for Saponins

The test solution was shaken with water. Copious lather formation indicated the presence of saponins.

2.10. Test for Steroids (Libber Mann Bur Chard Test)

To 2 ml of the test solution, a few drops of chloroform, 3-4 drops of acetic anhydride and a drop of concentrated sulphuric acid were added. Formation of purple colour that gradually changed to blue or green indicated the presence of steroids.

2.11. Test for Tannins

The test solution was mixed with basic lead acetate solution. Formation of the white precipitate indicated the presence of tannins.

2.12. Test for Terpenoids: (Noller Test)

The test solution was warmed with a piece of tin and a few drops of thionyl chloride. Formation of violet or purple colour indicated the presence of triterpenoids.

2.13. Test for Xanthoproteins

To the test solution, a few drops of concentrated nitric acid and a few drops of liquid ammonia were added. Formation of the reddish precipitate indicated the presence of xanthoproteins.

2.14. Test for Sugars

The test solution was mixed with equal volumes of Fehling's solution 'A' and 'B' and heated. Formation of brown colour indicated the presence of reducing sugars.

2.15. Anti-microbial Assay

2.15.1. Microbial Samples

The microorganisms were obtained from Department of Microbiology VHNSN College Virudhunagar, Tamil Nadu, India. The standard cultures were stored in slants at 4°C. Bacterial cell suspensions were prepared separately in nutrient broth by transferring a loopful of 24 hours grown culture from respective standard culture slants.

2.15.2. Determination of Bacterial Activity

The bactericidal activity of the crude extract was determined in accordance with the agar well diffusion method (Irobi *et al.*, 1994) 100 µl of cell suspension was spread on Mueller Hinton agar medium. Well were bored onto the agar using a sterile 6 mm diameter cork borer. 0.2 ml of the crude extract was added into the wells and incubated at 37°C for 24 h.The plates were observed for zone of inhibition after 24 h. These effects were compared with Tetracycline at a concentration of 10 μg/ml respectively. After incubation period, the zone of inhibition was measured by zone reader.

2.15.3. Determination of Antifungal Activity

The methanol extracts of the plant parts used were tested for their antifungal activity. The respective fungal spores were inoculated on the surface of Potato Dextrose Agar plates and incubated at 25°C for 3 days. Tetracycline, an antifungal compound at a concentration of 50 μg per well was used as control. After incubation period, the zone of inhibition was measured by zone reader.

3. Results and Discussion

In the present study, the phytochemical screening and antibacterial activities were performed with methanol extracts of the leaf of *Alstonia scholaris* R.Br and *Rauvolfia serpentina* (L.) Benth *ex* Kurz. The study was made against two gram negative bacteria using the standard disc diffusion method. The leaves of *Alstonia scholaris* methanol extract exhibited alkaloids, coumarins, phenols, tannins, saponins, terpenoids and quinones. And methanolic extract of *Rauvolfia serpentina* leaf revealed that alkaloids, anthroquinones, phenols, flavones, saponins, steroids and xanthoprotein. These phytochemicals confer antimicrobial activity on the leaf extracts (Table 1.1).

Table 1.1: Phytochemical screening of methanolic extract of *Alstonia scholaris* R.Br and *Rauvolfia serpentina* (L.) Benth.

Tests	Alstonias cholaris	Rauvolfia serpentina
Alkaloids	+	+
Anthroquinones	−	+
Coumarins	+	−
Phenols	+	+
Flavones	−	+
Tannins	+	−
Saponins	+	+
Steroids	−	+
Terpenoids	+	−
Xantho protein	−	+
Quinones	+	−
Sugars	−	−

(+) Presence of Phytochemical; (−) Absence of Phytochemical.

The results of the methanol leaves extract exhibited antibacterial activity against all the tested strain *viz.*, *Bacillus subtilis*, *Salmonella typhi* and *Aspergillus niger* fungus as shown in Table 1.2. The zones of inhibitions were produced by methanol extracts against all the test organisms. *Rauvolfia serpentina* methanol extracts were more active than the *Alstonia scholaris* methanol extract against all the microorganisms. The zones

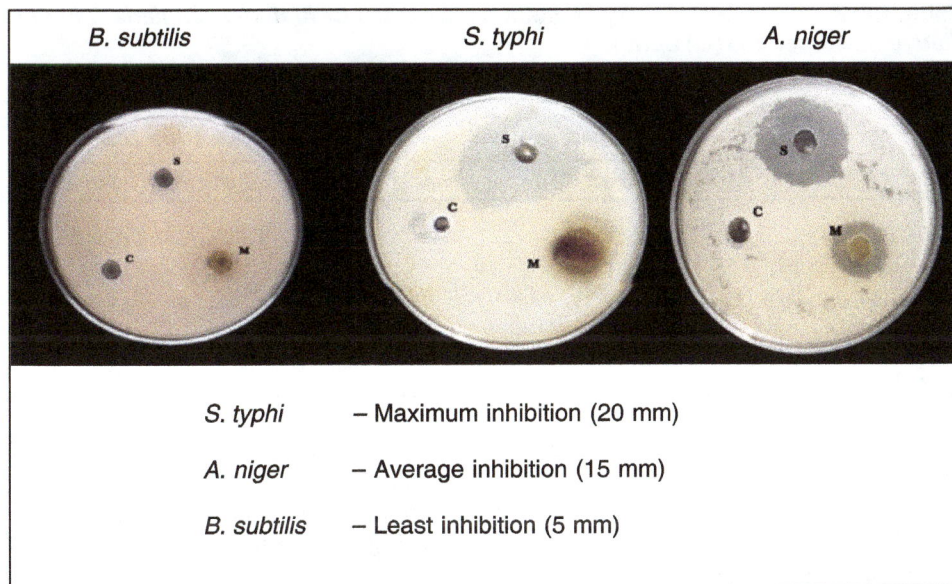

S. typhi	– Maximum inhibition (20 mm)
A. niger	– Average inhibition (15 mm)
B. subtilis	– Least inhibition (5 mm)

Figure 1.1: Antimicrobial activity of methanolic leaf extract of *Alstonia scholaris* (L.) R.Br.

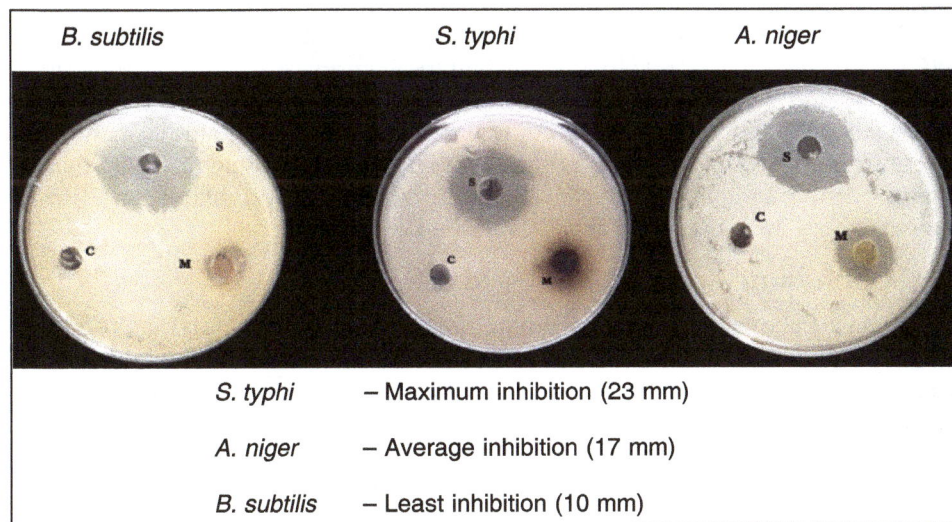

S. typhi	– Maximum inhibition (23 mm)
A. niger	– Average inhibition (17 mm)
B. subtilis	– Least inhibition (10 mm)

Figure 1.2: Antimicrobial activity of methanolic leaf extract of *Rauvolfia serpentina* Benth.

of inhibition were ranging from 20-23 mm in diameter. The highest zone of inhibitions (23 mm) noted in *Rauvolfia serpentina* methanol extract against *Salmonella typhi*. The fungus of *Aspergillus niger* exhibited 17 mm diameter.

Table 1.2: Antimicrobial activity of methanolic extract of *Alstonia scholaris* R.Br and *Rauvolfia serpentina* (L.) Benth.

Name of the Organism	Zone of Inhibition (mm)			
	Alstonia scholaris		Rauvolfia serpentina	
	Methanol Extract	Tetracycline	Methanol Extract	Tetracycline
Bacteria				
Bacillus subtilis	5	10	10	15
Salmonella typhi	20	40	23	43
Fungus				
Aspergillus niger	15	37	17	38

The various phytochemical compounds detected are known to have beneficial importance in medicinal sciences. For instance, flavonoids have been referred to as nature's biological response modifiers, because of their inherent ability to modify the body's reaction to allergies and virus and they showed their anti-allergic, anti-inflammatory, anti-microbial and anti-cancer activities (Aiyelaagbe and Osamudiamen, 2009). Plant steroids are known to be important for their cardiotonic activities and also possess insecticidal and antimicrobial properties. They are also used in nutrition, herbal medicine and cosmetics (Callow, 1936).

Tannins were reported to exhibit antiviral, antibacterial and anti-tumour activities. It was also reported that certain tannins were able to inhibit HIV replication selectively and was also used as diuretic (Callow, 1936). Saponin is used as mild detergents and in intracellular histochemical staining. It is also used to allow antibody access in intracellular proteins. In medicine, it is used in hypercholesterolemia, hyperglycemia, antioxidant, anticancer, anti-inflammatory, weight loss, etc. It is also known to have antifungal properties (Haslem, 1989).

4. Conclusion

From the above studies it is concluded that the traditional plants may represent new sources of antimicrobial with stable, biologically active components that can establish a scientific base for the use of plant in modern medicine. The local ethno medical preparations and prescriptions of plant sources should be scientifically evaluated and then disseminated properly and the knowledge about the botanical preparation of traditional sources of medicinal plants can be extended for future investigation into the field of pharmacology, photochemistry, ethno botany and other biological actions for drug discovery.

References

Abbas, J.A., Ahmed. A., and Mahasneh. A.M., 1992. Herbal plants in the traditional medicine of Bahrain. *Eco. Bot.*, 46: 158–163.

Adamu, M., Alibilar. Monday odis and chelae match awe., 200 Antimicrobial activity and phytochemical screening of some selected medicinal plants in Bauchi. *J. Eco. Taxon. Bot.*, 24: 123–127.

Aiyelaagbe, O.O. and Osamudiamen, P.M., 2009. Phytochemical screening for active compounds in *Mangifera indica. Plant, Sci. Res.*, 2(1): 11–13.

Anonymous, 1966. The wealth of India a dictionary of Indian raw materials and industrial products. Vol. VII. Publications and Information Directorate. C.S.I.R. New Delhi, pp. 76 and 77.

Bedi, S.I., 1978. Ethnobotany of the Raton mahal hills, Gujarat, India. *Eco. Bot.*, 32: 278–284.

Bhakuni, D.S., 1990. Drugs from plants. *Sci. Rep.*, 18: 12–17.

Callow, R.K., 1936. Steroids. *Proc. Royal, Soc. London* Series, 157, 194.

Chopra, R.N., Nayar, S.L. and Chopra, I.C., 1956. *Glossary of Indian Medicinal Plants.* CSIR, New Delhi.

Dubey, N.K., Tripathi, Pramila and Singh, H.B., 2000. Prospects oil as antifungal agents. *J. Med. and Plant Sci.*, 22: 350–354.

Haslem, E., 1989. *Plant Polyphenols: Vegetable Tannins Revised–Chemistry and Pharmacology of Natural Products.* Cambridge University Press, p. 169.

Holdsworth, D., 1986. *Medicinal Plants of Papua New Guinea.*

Jain, S.K., 1963. Studies Indian ethnobotany.Less known uses of fifty common plants from the tribal areas of Madhya Pradesh. *Bull. Bot. Surv., India*, 5: 223–226.

Jain, S.K., Benerjee, D.K. and Pal, D.C., 1973. Medicinal plants among certain Ad basis in India. *Bull. Bot. Surv., India*, 15: 85–91.

Jain, S.K. and Dam, N., 1979. Some ethno-botanical notes from northeastern India. *Eco. Bot.*, 33: 52–56.

Joshi, P., 1982. An ethnobotanocal study of Bhills a preliminary survey. *J. Econ. Tax. Bot.*, 3: 257–266.

Joshi, P., 1989. Herbal drugs in tribal Rajasthan–from child birth to childcare. *Ethnobotany*, 1: 77–87.

Knobloch, K., Weis, N. and Weigand, 1986. Mechanism of antimicrobial activity of essential oils. *Planta Medica*, pp. 552–556.

Lala, P.K., 1993. *Lab Manuals of Pharmacognosy*, 5th edn. CSI Publishers and Distributors, Kolkata.

Lalitha Kumari, H. and Sirsi, M., 1965. Antibacterial and antifungal activities of *Areca catechu* L. *Indian. J. Exp. Boil*, 3: 75–82.

Nayar, M.P. and Sastry, A.R.K., 1987. *Red Data Book of Indian Plants, Vols. I, II and III.* Published by the Director, Botanical Survey of India, Kolkata.

Perumal Samy, R., Ignacimuthu, S. and Sen. A., 1998. Screening of 34 Indian medicinal plants for antibacterial properties. *J. Ethnopharm.*, 62: 173–182.

Prasad, P.N., Jabadhas, A.W. and Janaki Ammal, E.K., 1987. Medicinal plants used by the Kanikkars of South India. *J. Econ. Tax. Bot.*, 11: 149–155.

Pushpangadan, P. and Atal, C.K., 1984. Ethno-medico-botanical investigations in Kerala. I. Some primitive tribals of Western Ghats and their herbal medicine. *J. Ethno-pharmacology*, 11: 59–77.

Pushpangadan, P. and Atal, C.K., 1986. Ethno-medical and ethnobotanical investigations among some scheduled caste communities of Travancore, Kerala, India. *J. Ethno-pharmacology*, 16: 175–190.

Rama Chandran, V.S. and Nair, N.C., 1981. Ethnobotanical observations in Irulars of Tamil Nadu (India). *J. Econ. Tax. Bot.*, 2: 183–190.

Sharma, R.N. and Saxena, V.K., 1996. *In vitro* antimicrobial efficacy of leaves extract of *Bauhinia varigata*. *Asian T. Chem.*, 8: 811–812.

Perry, L.M. and Metzger, H., 1980. *Medicinal Plants of East and Southeast Asia*. MIT Press, Cambridge.

Zashim Uddin, M., Hassan M.A. and Sultan M. (2006). Ethnobotanical survey of medicinal plants in Phulbari Upazila of Dinajpur district, Bangladesh. *Bangladesh J. Plant Taxon.*, 13(1): 63–68.

Scientific Basis of Herbal Medicine (2013)   *Pages 9–30*
Editor: **Dr. Parimelazhagan Thangaraj**
Published by: **DAYA PUBLISHING HOUSE, NEW DELHI**

Chapter 2

Preliminary Phytochemical Screening and *In vitro* Anticancer Activity of the Weed Plant *Sphaeranthus indicus* Linn.

**B. Thamilmaraiselvi[1], G. Theivandran[2], M. Ranjani[1],
H. Deena Priscilla[1] and S. Sindhu[1]**

*[1]Department of Microbiology, Cauvery College for Women,
Tiruchirappalli – 620 018, Tamil Nadu, India
[2]Department of Zoology, VOC College, Millerpuram,
Tuticorin – 628 008, Tamil Nadu, India*

1. Introduction

The use of plants and plant products; as medicine can be traced as far back as the beginning of human civilization. Medicinal plants are the richest bio-resource of drugs of traditional system of medicine, modern medicine, nutraceuticals food supplements, folk medicines, pharmaceutical intermediate and lead compounds in synthetic drugs (Ncube *et al.*, 2008). Phytochemicals are non-nutritive plant chemicals that are more complex and specific and exert their action by resembling endogenous metabolites. Ligands, hormones, signal transduction molecules or neurotransmitters due to similarities in their target sites. These plant based natural constituents can be derived from any part of the plant like bark, leaves, flowers, roots, fruit and seeds, etc. According to World Health Organization (WHO), more than 80 per cent of the world's population relies on traditional medicines for their primary healthcare needs. Contrary to the synthetic drugs, antimicrobials of plant origin are not associated with many side effects and have an enormous therapeutic potential to heal many infectious diseases. Therefore, researchers are increasingly turning their attention to folk medicine, looking for new leads to develop better drugs against microbial infections.

Sphaeranthus indicus Linn is one of the six species of the genus *Sphaeranthus* from the family compositae (Asteraceae). The herb is used as a laxative, digestible, demulcent, emollient and pectoral. All parts of the plant find medicinal uses. The whole herb is used in Ayurvedic preparations to treat epilepsy, mental disorders and hemicranins (Ambavade *et al.,* 2006 and Jha *et al.,* 2010). Hot water extract of the entire plant is used for glandular swelling of the neck and for jaundice (Ikram, 1981).

The plant appears to have a broad spectrum of activity on several ailments various parts of the plant have been explored for anxiolytic activity, neuroleptic activity, immunomodulatory activity, anti-inflammatory activity, antihyperglycemic activity, hepatoprotective activity, antihyperlipidemic activity and many other miscellaneous activities. It is reported to contain eudesmanoids, eudesmandides, sesquiterpene lactones, flavanoids and sterop glycoside. However, less information is available regarding the clinical, toxicity and phytoanalytical properties of the plant. In the present study, the phytochemical and *in vitro* anticancer properties of the weed plant *S. indicus* were eviduated for its therapeutic preparations.

2. Materials and Methods

2.1. Collection of Plant Materials and Extraction

The plant material was collected from the rice fields of Manikandam, Tiruchirappalli district, Tamil Nadu in the month of January, 2009. The herbarium specimen was deposited in Rapinat Herbarium, St. Joseph's College, Trichy, Tamil Nadu, India.

The collected plant was washed thoroughly in tap water, shade dried and leaves, flower stems and roots of the plant were separated and coarsely powdered by a mechanical grinder. About 10 gm of each dried powder of the plant were soaked separately in 100 ml of different solvents like methanol, ethanol, and chloroform and petroleum ether in conical flasks and then subjected to agitation on a rotary shaker for three days at 190-220 rpm. After 72 hrs. The plant extracts were filtered through Whatmann no. 42 paper separately and they were allowed to be concentrated by heating on a boiling water bath. Concentrated extracts were then preserved in labelled sterilized air tight bottles at 4°C in refrigerator until when required for use. Aqueous extract of the plant was prepared by soaking 10 gm of dried powder in 100 ml of boiling water and the extract was also collected as described earlier.

Leaves, flower, stems and roots extract of different solvents were analysed for the presence of alkaloids, saponins, tannins, phenols, steriods, cardiac glycosides, anthraqunions, flavonoids, terpernoids, aminoacids, reducing sugars and monosaccharides as described by Soffowara (1994) and Siddiqui and Ali (1997).

2.2. Quantitative Estimation of Phytochemicals

2.2.1. Alkaloid Determination (Harborne, 1973)

5 g of the sample was weighed into a 250 ml beaker and 200 ml of 10 per cent acetic acid in ethanol was added and covered and allowed to stand for 4 h. This was filtered and the extract was concentrated on a water bath to one-quarter of the original volume. Concentrated ammonium hydroxide was added dropwise to the extract until the precipitation was complete. The whole solution was allowed to settle and

the precipitated was collected and washed with dilute ammonium hydroxide and then filtered. The residue is the alkaloid, which was dried and weighed.

2.2.2. Flavonoid Determination (Bohm and Kocipai-Abyazan, 1994)

10 g of the plant sample was extracted repeatedly with 100 ml of 80 per cent aqueous methanol at room temperature. The whole solution was filtered through whatman filter paper No. 42 (125 mm). The filtrate was later transferred into a crucible and evaporated into dryness over a water bath and weighed.

2.2.3. Saponin Determination (Obadoni and Ochuko, 2001)

The samples were ground and 20 g of each were put into a conical flask and 100 cm^3 of 20 per cent aqueous ethanol were added. The samples were heated over a hot water bath for 4 h with continuous stirring at about 55°C. The mixture was filtered and the residue re-extracted with another 200 ml 20 per cent ethanol. The combined extracts were reduced to 40 ml over water bath at about 90°C. The concentrate was transferred into a 250 ml separating funnel and 20 ml of diethyl ether was added and shaken vigorously.

The aqueous layer was recovered while the ether layer was discarded. The purification process was repeated. 60 ml of n-butanol was added. The combined n butanol extracts were washed twice with 10 ml of 5 per cent aqueous sodium chloride. The remaining solution was heated in a water bath. After evaporation the samples were dried in the oven to a constant weight; the saponin content was calculated as percentage.

2.2.4. Determination of Total Phenols (Harborne, 1973)

The fat free sample was boiled with 50 ml of ether for the extraction of the phenolic component for 15 min. 5 ml of the extract was pipetted into a 50 ml flask, then 10 ml of distilled water was added. 2 ml of ammonium hydroxide solution and 5 ml of concentrated amylalcohol were also added. The samples were made up to mark and left to react for 30 min for colour development. This was measured at 505 nm.

2.2.5. Tannin Determination (Van-Burden and Robinson, 1981)

500 mg of the sample was weighed into a 50 ml plastic bottle. 50 ml of distilled water was added and shaken for 1 h in a mechanical shaker. This was filtered into a 50 ml volumetric flask and made up to the mark. Then 5 ml of the filtrate was pipetted out into a test tube and mixed with 2 ml of 0.1 M FeCl$_3$ in 0.1 N HCl and 0.008 M potassium ferrocyanide. The absorbance was measured at 120 nm within 10 min.

2.3. HPTLC Fingerprinting of Methanolic Leaves Extract

Plant extracts were dissolved in respective solvents. Sample was applied on 0.22mm thick silica gel plate by making use of Camag automatic TLC sampler. Sample was applied as band and is not spot. Chromatogram was developed with active stationary, mobile and vapour phases. Stationary phase plate was put into the mobile phase containing organic solvents like hexane: ethyl acetate: methanol in the ratio of 1: 2: 1. Stepwise automatic procedure was followed at room temperature to run the column. Automatic developing chamber was used to develop the chromatogram. Chemical compounds are quantitatively and evaluated through spectral scanner. Scanning was controlled by Camag software © 1998 available in the instrument. Computerized scanning HPTLC report provided the informations like Rf value, λ max and per cent of chemical constituents present in the sample. The results were recorded and interpreted.

2.4. Identification of Compounds using GC-MS

GC-MS analysis was carried out by using Perkin Elmer-Clarus 500 GC-MS Unit. The analysis was carried out to detect the possible compounds present in the active fraction of the column type used was PE-S (equivalent to DB-5) with a column length of 30 m using carrier gas as Helium. The flow rate maintained was 1 ml/min with an initial column temperature of 50°C and final temperature of 250°C. The rate of temperature change in the column was maintained as 5°C/min. 1 M volume of sample was taken for injection. The identification of compounds was accomplished using computer searches in commercial libraries.

2.5. *In vitro* Anticancer Activity Studies

2.5.1. *In vitro* Cytotoxicity by MTT Assay Method (Mossman, 1983)

The methanolic and aqueous leaves and flower extracts were subjected to *in vitro* cytotoxicity by MTT assay method using HEP-G2 cell lines (Human Liver Cancer cell lines).

2.5.2. *In vitro* Cytotoxicity by Short-Term Dye Exclusion Method (Freshny, 1994)

The methanolic and aqueous leaf and flower extracts were subjected to tryphan blue exclusion assay on HEP-G2 cell line.

3. Results

All the parts of the plant *S. indicus* (leaves, flowers stem and root) were subjected to preliminary phytochemical analysis. Tables 2.1 and 2.2 showed the presence of secondary metabolites such as alkaloids, saponins, flavanoids, tannins, and phenolic compounds. Of the four plant parts, the methanolic and aqueous extract of *S. indicus* leaves showed the presence of saponins, phenolic compounds, tannins, flavanoids and alkaloids. Since most of phytoconstituents were found in methanolic and aqueous extracts of *S. indicus* leaves and flowers. Hence these two plants parts were subjected to quantitative analysis.

Quantitative estimation of crude extracts of plant showed that, the methanolic and aqueous extract of *S. indicus* leaves were rich in alkaloids, tannins, saponins, phenolic compounds, and flavanoids (Table 2.3) whereas the flower extracts contain only moderate quantity of the above phytochemicals (Table 2.4).

Table 2.1: Phytochemical screening of *Sphaeranthus indicus* leaves extracted with different solvents.

Phytochemicals	Methanolic Extract	Aqueous Extract	Ethanolic Extract	Chloroform Extract	Petether Extract
Alkaloids	++	+	+	–	–
Steroids	+	–	+	–	+
Terpenoids	++	–	–	–	–
Flavanoids	++	++	+	–	–
Saponins	+++	++	+	–	–
Phenolic compounds	+++	++	+	–	–
Tannins	+++	++	+	–	–
Cardiac glycosides	–	–	–	+	–
Amino acids	++	+++	+	+	–
Proteins	++	+	+	–	–
Mono saccharides	++	+	+	–	–
Reducing sugar	+++	+	+	–	–

+++: Strongly present; ++: Moderately present; +: Poorly present; –: Absent.

Table 2.2: Phytochemical screening of *Sphaeranthus indicus* flower extracted with different solvents.

Phytochemicals	Methanolic Extract	Aqueous Extract	Ethanolic Extract	Chloroform Extract	Petether Extract
Alkaloids	–	+	–	–	–
Steroids	+	–	+	–	+
Terpenoids	–	+	–	+	–
Flavanoids	+	+	–	–	–
Saponins	+	+	–	–	–
Phenolic compounds	++	++	+	–	–
Tannins	++	++	+	–	–
Cardiac glycosides	–	–	–	–	–
Amino acids	++	+	–	–	–
Proteins	+++	+	+	–	–
Mono saccharides	+++	+	+	–	–
Reducing sugar	+++	+	+	–	–

+++: Strongly present; ++: Moderately present; +: Poorly present; –: Absent.

3.1. GC-MS Analysis of the Plant Extract

GC-MS analysis was carried out to detect the possible compounds present in the active fraction. The methanolic leaves extracts of *S. indicus* was analysed using GC-MS. Nearly 115 compounds were identified. The compound name, retention time, peak area, and percentage peak area and structure areas identified and are recorded. The spectrum of unknown compound was compared with the spectrum of the known compound stored in the NIST library.

Table 2.3: Quantitative estimation of important phytochemical in the crude extracts of S. indicus leaves.

Extracts	Alkaloids (mg/g)	Phenols (mg/g)	Tannins (mg/g)	Flavanoids (mg/g)	Saponins (mg/g)
Methanolic extract	2.91±0.03	1.87±0.12	13.26±1.10	2.62±0.12	2.62±0.13
Aqueous extracts	1.62±0.12	1.23±0.28	9.12±0.20	0.55±0.18	1.27±0.28
Ethanolic extracts	0.32±0.12	0.51±0.14	3.45±0.23	0.47±0.11	1.12±0.28
Chloroform extract	0.00	0.00	0.00	0.00	0.00
Pet Ether extract	0.00	0.00	0.00	0.00	0.00

Table 2.4: Quantitative estimation of important phytochemical in the crude extracts of S. indicus flowers.

Extracts	Alkaloids (mg/g)	Phenols (mg/g)	Tannins (mg/g)	Flavanoids (mg/g)	Saponins (mg/g)
Methanolic extract	0.97±0.11	1.71±0.02	9.23±0.24	0.49±0.21	1.19±0.13
Aqueous extracts	0.72±0.11	0.81±0.11	5.49±0.21	0.69±0.12	1.71±0.52
Ethanolic extracts	0.00	1.13±0.03	6.11±1.11	0.00	0.00
Chloroform extract	0.00	0.00	0.00	0.00	0.00
Pet Ether extract	0.00	0.00	0.00	0.00	0.00

3.2. HPTLC Finger Printing Analysis of Methanolic Leaves Extract

The densitometric HPTLC analysis of methanolic leaf extract was performed for development of characteristic finger printing profile. The R_f value of methanolic leaves extract (10 µl) was found to be 0.34, which is closer to 0.35. According to the literature reference, R_f value at 0.35 is a sesquiterpene glycoside (sphaeranthanolide). The mobile phase used was dichloromethane methanol: acetic acid (6.5: 3.0: 0.5) (Plate 2.1).

3.3. *In vitro* Cytotoxicity against Human Liver Cell Line (HePG2) by MTT Assay Method

The cytotoxic effect of the plant extract on HePG2 cell line was determined by MTT assay. The ability of the cell to survive a toxic insult has been the basis of cytotoxicity assays and the potential cytotoxic value was calculated by determining the toxic concentration which may kill upto 50 per cent of the total cell used (IC_{50}).

The methanolic leaves extract displayed strongest cytotoxic effect on human liver cancer cell line (HePG2) with IC_{50} value of 15 µg/ml (Table 2.5). The aqueous leaves extract shows the cytotoxic effect on human liver cancer cell line with IC_{50}

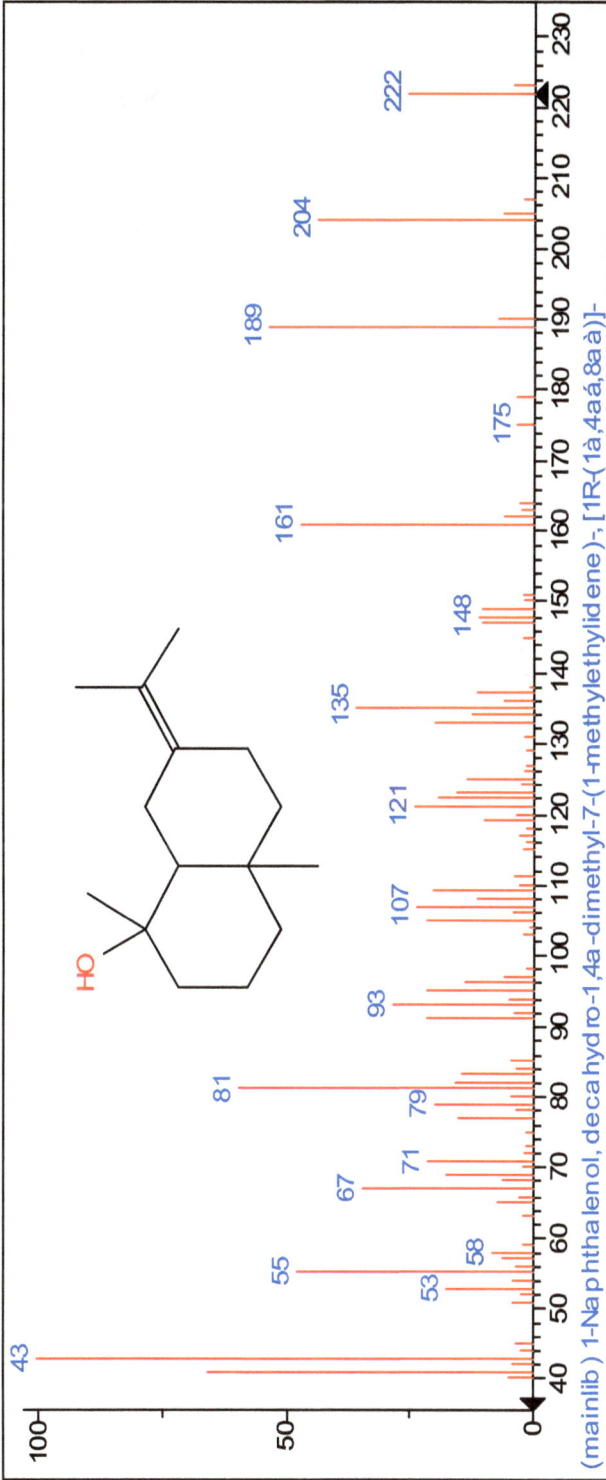

(mainlib) 1-Naphthalenol, decahydro-1,4a-dimethyl-7-(1-methylethylidene)-, [1R-(1à,4aá,8aà)]-

Figure 2.1a: GC-MS Profile.

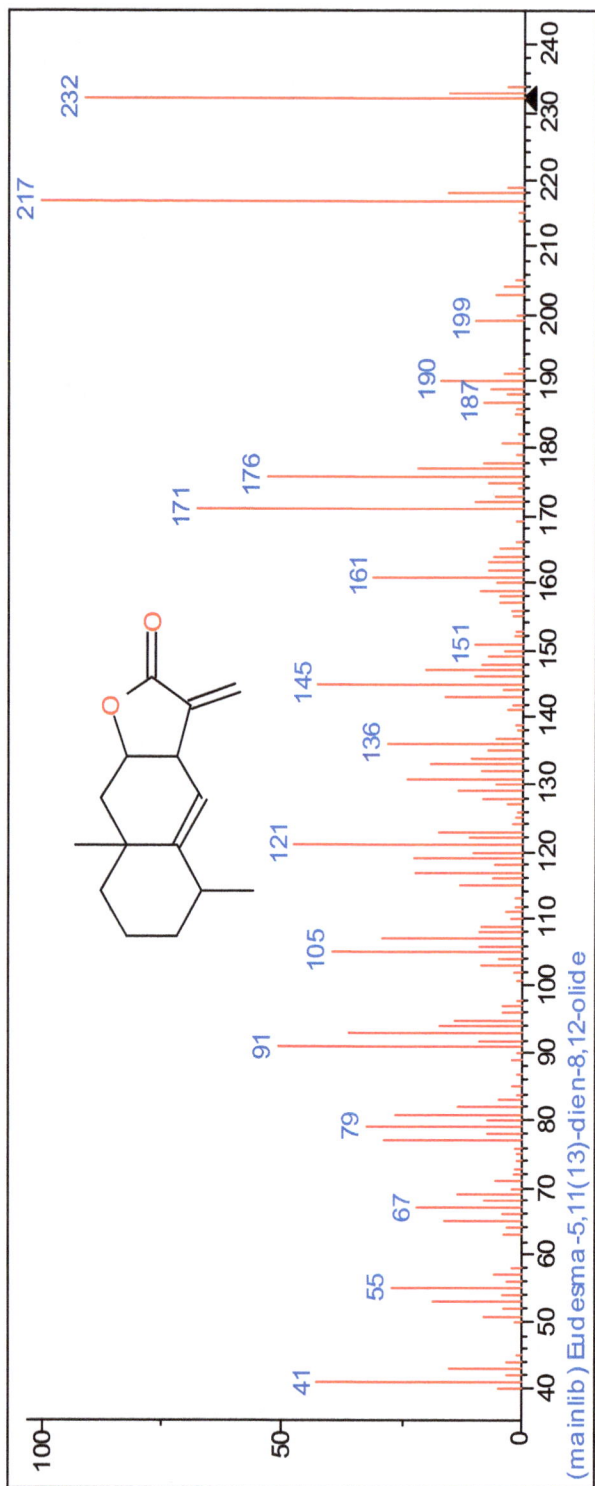

(mainlib) Eudesma-5,11(13)-dien-8,12-olide

Figure 2.1b: GC-MS Profile.

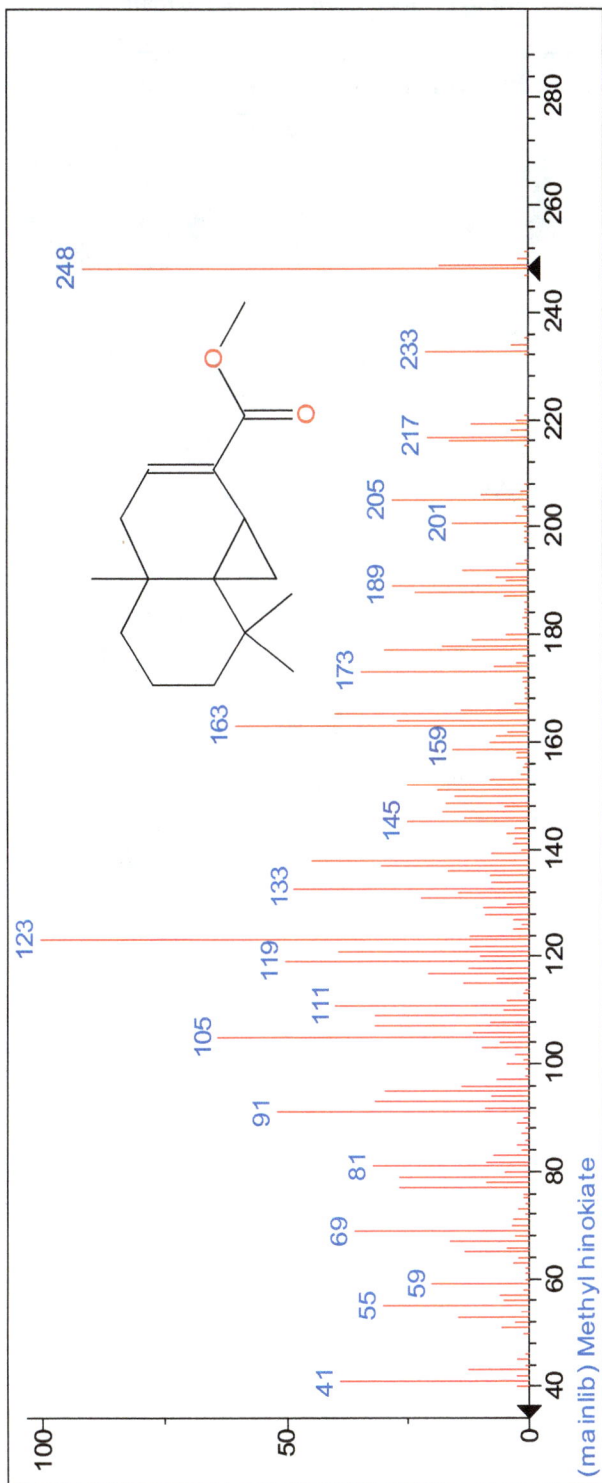

(mainlib) Methyl hinokiate

Figure 2.1c: GC-MS Profile.

Plate 2.1: HPTLC fingerprinting of *Sphaeranthus indicus* methanolic leaves extract.

A. HPTLC under 254 nm
(mobile phase dichlromethane: methanol: acetic acid (6.5: 2.0: 0.5)

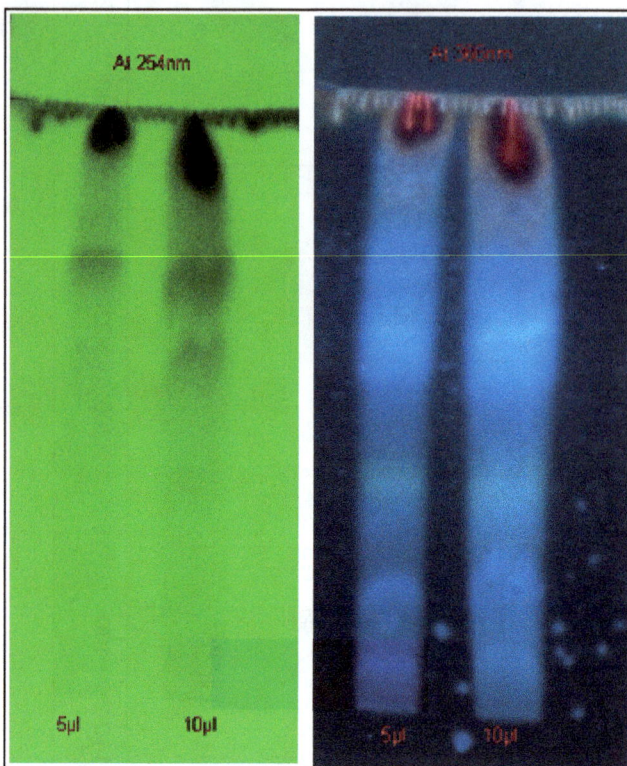

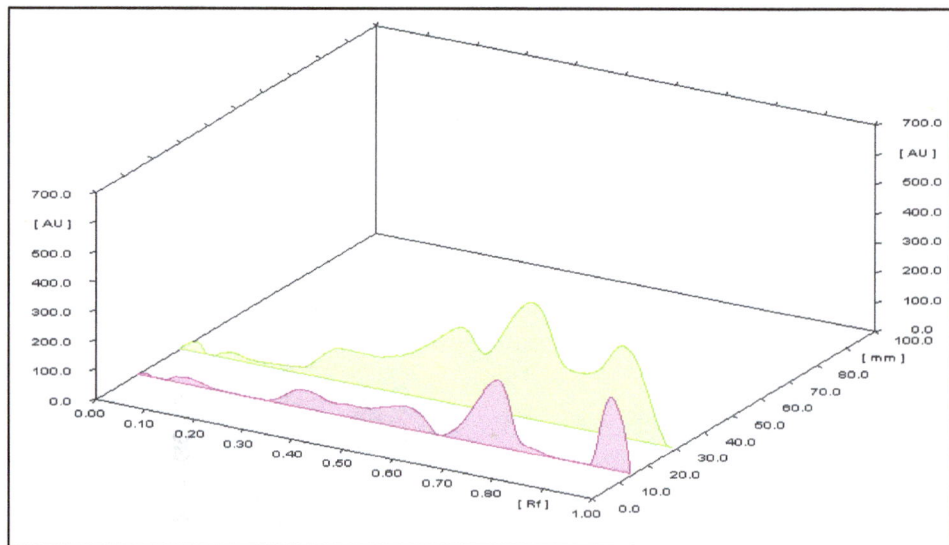

Three Dimensional view

a. HPTLC Methanolic leaves extract (10 µl)–peak display and peak table.

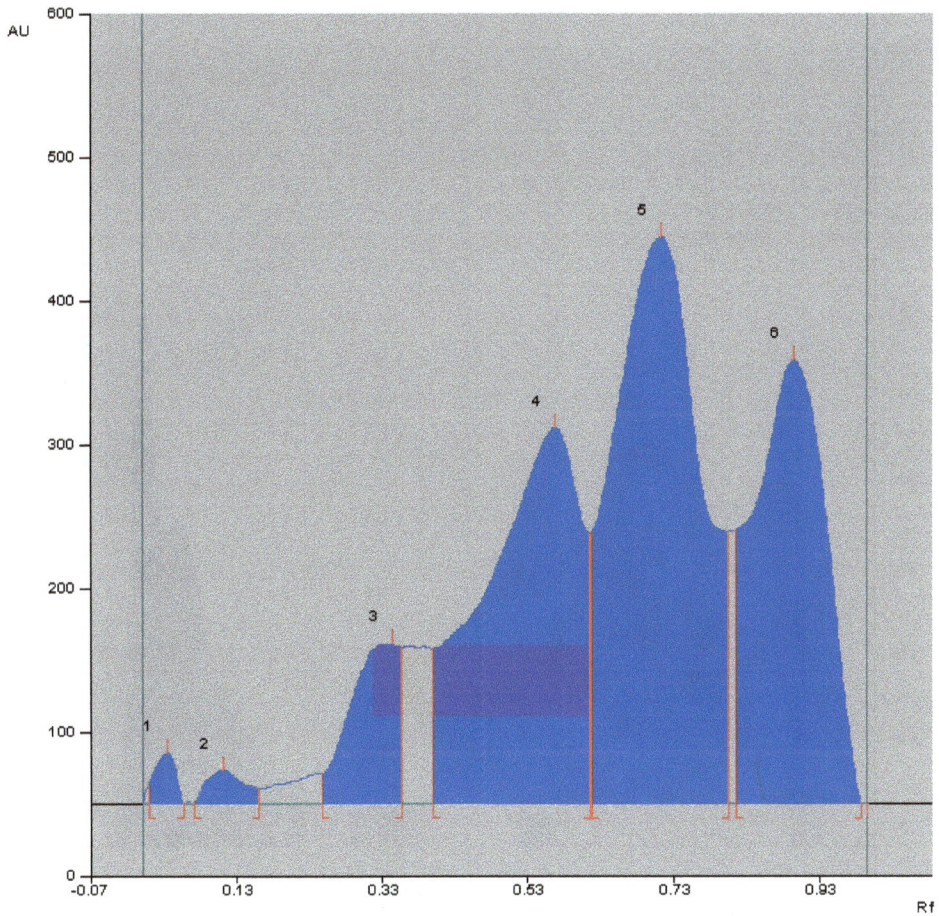

Track 2 . ID: Sphaeranthus-Alcoholic

Peak	Start Rf	Start Height	Max Rf	Max Height	Height %	End Rf	End Height	Area	Area %
1	0.01	12.3	0.03	34.7	3.06	0.06	0.4	796.9	0.82
2	0.07	0.6	0.11	22.9	2.02	0.16	10.0	954.8	0.98
3	0.24	20.7	0.34	110.9	9.79	0.36	108.4	6024.1	6.20
4	0.40	107.5	0.57	261.3	23.08	0.61	188.1	27474.7	28.30
5	0.62	188.3	0.71	394.1	34.80	0.80	188.9	37194.9	38.31
6	0.81	189.8	0.89	308.6	27.25	0.99	0.4	24643.4	25.38

b. HPTLC Methanolic leaves extract (5 µl)–peak display and peak table.

Track 1 , ID: Sphaeranthus-Alcoholic

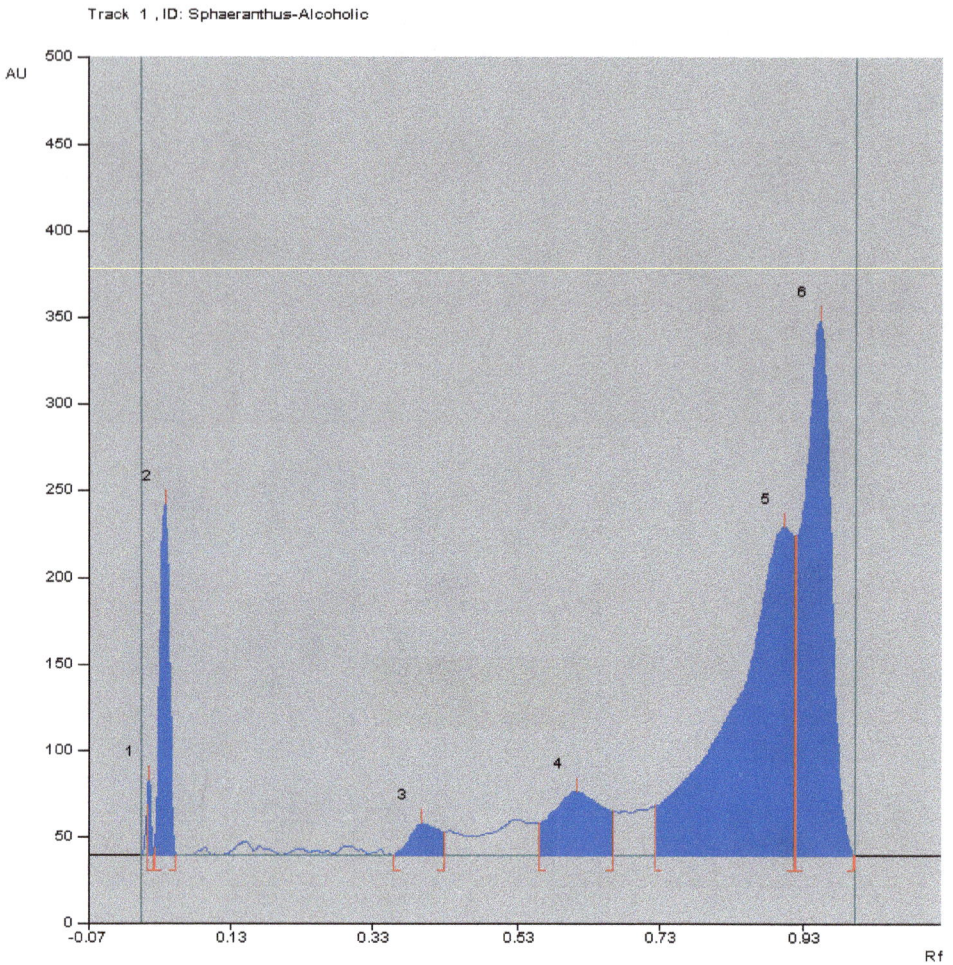

Peak	Start Rf	Start Height	Max Rf	Max Height	Height %	End Rf	End Height	Area	Area %
1	0.01	29.3	0.01	44.0	5.49	0.02	0.0	184.6	0.70
2	0.02	4.2	0.03	203.1	25.34	0.05	0.0	2062.1	7.85
3	0.36	0.2	0.39	18.6	2.31	0.43	13.3	575.8	2.19
4	0.56	18.6	0.61	37.1	4.63	0.66	25.5	2068.9	7.88
5	0.72	28.9	0.90	189.7	23.67	0.91	184.4	12331.8	46.97
6	0.92	184.7	0.95	309.1	38.57	1.00	1.0	9033.8	34.41

c. HPTLC Methanolic flower extract (10 µl)–peak display and peak table.

Track 2 , ID: Sphaeranthus-Alcoholic

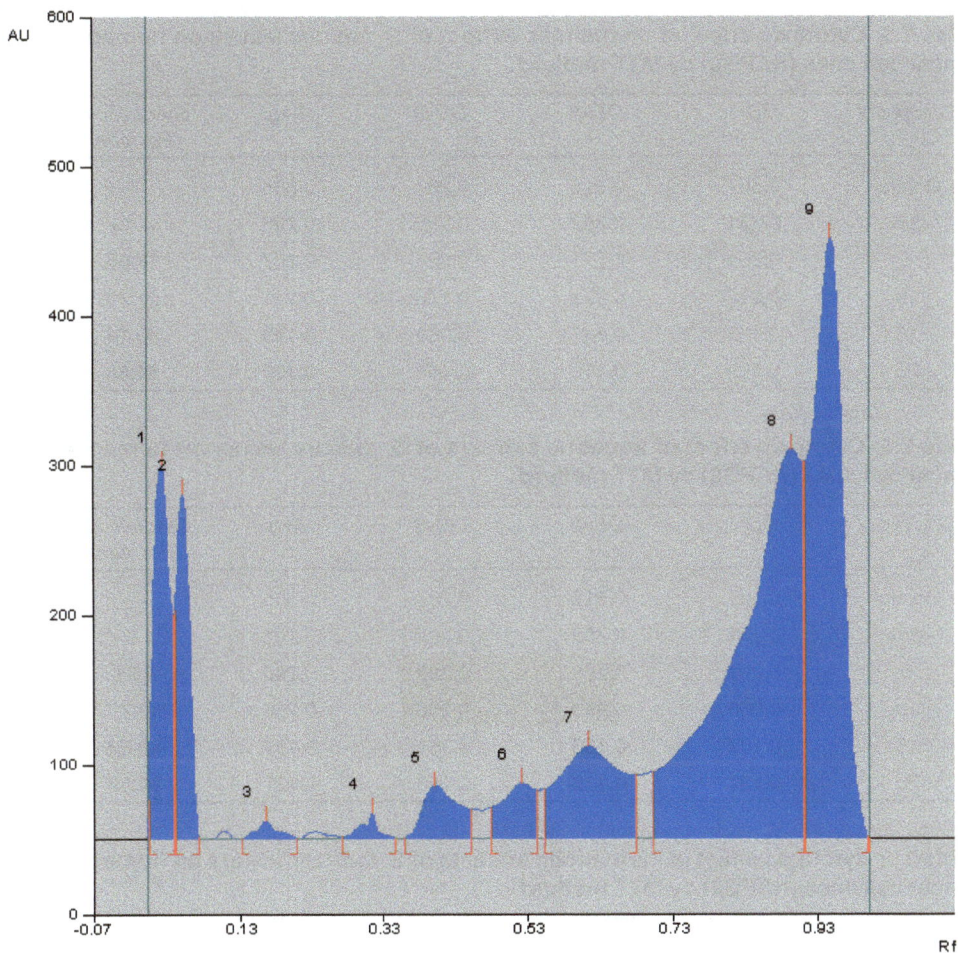

Peak	Start Rf	Start Height	Max Rf	Max Height	Height %	End Rf	End Height	Area	Area %
1	0.00	25.3	0.02	249.3	19.10	0.04	149.8	4284.1	9.20
2	0.04	152.6	0.05	231.4	17.73	0.07	0.7	3207.3	6.89
3	0.13	0.3	0.16	11.9	0.91	0.21	0.2	260.2	0.56
4	0.27	0.7	0.31	17.8	1.36	0.34	0.4	278.3	0.60
5	0.36	0.1	0.40	35.4	2.71	0.45	19.1	1377.8	2.96
6	0.48	21.0	0.52	36.8	2.82	0.54	32.3	1328.8	2.85
7	0.55	33.2	0.61	61.4	4.71	0.67	42.0	4069.1	8.74
8	0.70	44.1	0.89	259.9	19.91	0.91	251.2	18487.5	39.70
9	0.91	251.3	0.94	401.4	30.75	1.00	1.2	13269.6	28.50

value of 20 µg/ml (Table 2.6). The inhibition of cell growth by methanolic and aqueous flower extracts was found to be less than that mediated by leaves extract (Tables 2.7 and 2.8).

Table 2.5: Cytotoxic effect of methanolic extract of S. indicus leaves on human liver cancer cell lines (HEPG2) by MTT method.

Con µg/ml	OD-1	OD-2	OD-3	Avg	Cytotoxicity (Per cent)
Control	0.307	0.312	0.297	0.305	—
2.5	0.291	0.302	0.285	0.293	4.15
5	0.261	0.271	0.268	0.267	12.66
10	0.217	0.221	0.226	0.221	27.51
15	0.143	0.147	0.139	0.143	53.14
25	0.109	0.113	0.105	0.109	**64.26**

Table 2.6: Cytotoxic effect of aqueous extracts of S. indicus leaves on human liver cancer cell lines (HEPG2) by MTT method.

Con µg/ml	OD-1	OD-2	OD-3	Avg	Cytotoxicity (Per cent)
Control	0.305	0.312	0.297	0.305	—
2.5	0.301	0.309	0.294	0.301	1.09
5	0.294	0.288	0.299	0.290	4.91
10	0.260	0.263	0.266	0.266	13.77
15	0.157	0.163	0.151	0.157	48.52
25	0.121	0.126	0.116	0.121	**60.32**

Table 2.7: Cytotoxic effect of methanolic extracts of S. indicus flowers on human liver cancer cell lines (HEPG2) by MTT method.

Con µg/ml	OD-1	OD-2	OD-3	Avg	Cytotoxicity (Per cent)
Control	0.278	0.277	0.283	0.279	—
2.5	0.271	0.273	0.269	0.271	2.86
5	0.241	0.239	0.243	0.241	13.62
10	0.236	0.238	0.234	0.236	15.41
15	0.209	0.211	0.207	0.209	25.08
25	0.186	0.189	0.183	0.183	**34.40**

3.4. *In vitro* Cytotoxicity by Short Term due Exclusion Method

The methanolic and aqueous extracts of S. indicus leaves, flower were subjected to tryphan blue dye exclusion method (Tables 2.9–2.12). The various concentration of the plant extract used are 100, 250, 500 and 100 µg/ml in the cell count was observed with the increased concentration of the extract. The methanolic and aqueous leaves extract of *S. indicus* showed 82.20 per cent and 70.94 per cent reduction in cell count at 1000 µg/ml concentration (Tables 2.9 and 2.10; Plate 2.2). Whereas the methanolic and aqueous flower extract of *S. indicus* yielded 66.67 per cent and 54.36 per cent dead cells at 1000 µg/ml concentration respectively (Tables 2.11 and 2.12; Plate 2.3).

Table 2.8: Cytotoxic effect of aqueous extracts of *S. indicus* flowers on human liver cancer cell lines (HEPG2) by MTT method.

Con µg/ml	OD-1	OD-2	OD-3	Avg	Cytotoxicity (Per cent)
Control	0.278	0.277	0.283	0.279	—
2.5	0.277	0.279	0.281	0.277	0.95
5	0.245	0.251	0.239	0.245	12.18
10	0.239	0.241	0.237	0.239	14.33
15	0.223	0.228	0.231	0.227	18.62
25	0.193	0.187	0.198	0.193	**31.03**

Table 2.9: Cytotoxic effect of methanolic extracts of *S. indicus* leaves on human liver cancer cell lines (HEPG2) by Tryphan blue dye exclusion method.

Sl.No.	Conc. of Plant Extract (µg/ml)	No. of Viable Cells	Viable Cells (per cent)	No. Dead Cells	Dead Cells (per cent)
1.	Control	114	95	6	5
2.	100	54	62.07	33	37.93
3.	250	53	46.91	60	53.09
4.	500	38	31.41	83	68.59
5.	1000	21	17.80	97	**82.20**

Table 2.10: Cytotoxic effect of aqueous extracts of *S. indicus* leaves on human liver cancer cell lines (HEPG2) by Tryphan blue dye exclusion method.

Sl.No.	Conc. of Plant Extract (µg/ml)	No. of Viable Cells	Viable Cells (per cent)	No. Dead Cells	Dead Cells (per cent)
1.	Control	113	96.59	4	3.41
2.	100	137	72.30	52	27.69
3.	250	81	56.25	63	43.75
4.	500	64	43.84	82	56.16
5.	1000	43	29.06	105	**70.94**

Plate 2.2: Cytotoxic nature of *Sphaeranthus indicus* leaves extract.

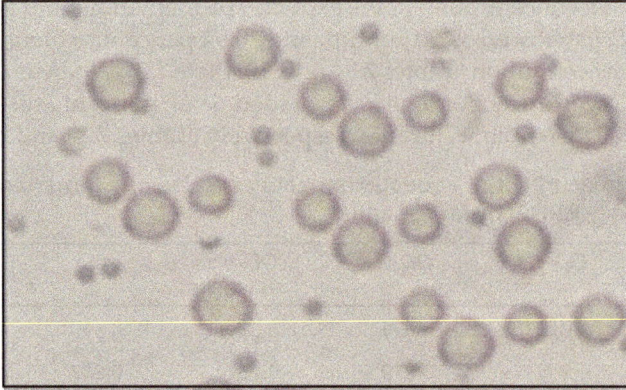

Normal HEPG2 cells

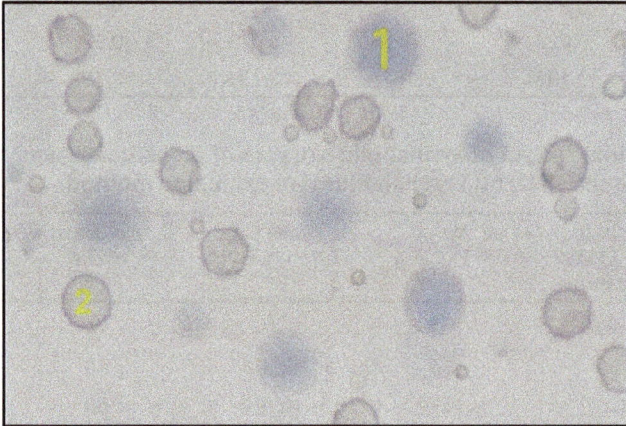

Cytotoxic nature of methanolic extract of *Sphaeranthus indicus* leaves (1) Blue–Dead cells; (2) White–Live cells

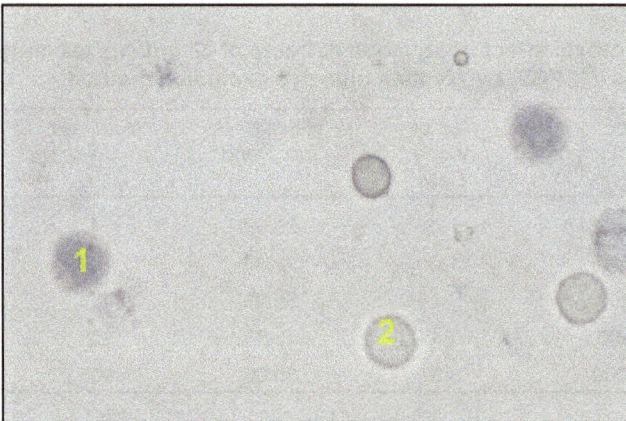

Cytotoxic nature of aqueous extract of *Sphaeranthus indicus* leaves (1) Blue–Dead cells; (2) White–Live cells

Plate 2.3: Cytotoxic nature of *Sphaeranthus indicus* flower extract.

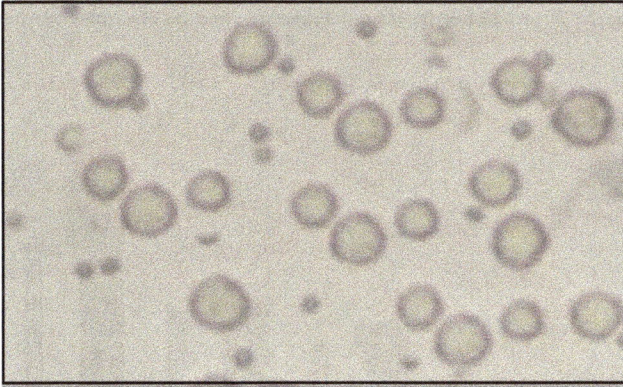

Normal HEPG2 cells

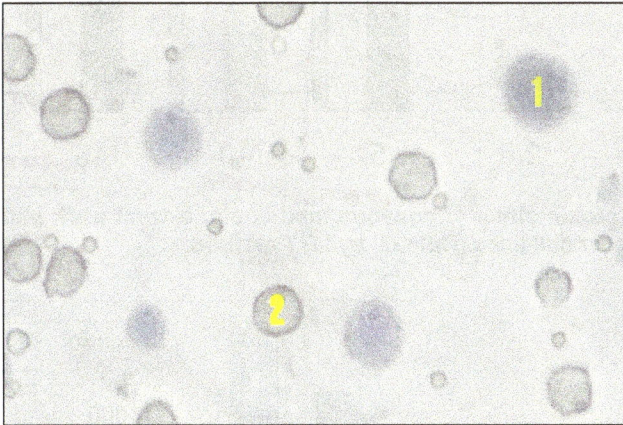

Cytotoxic nature of methanolic extract of *Sphaeranthus indicus* flower
(1) Blue–Dead cells; (2) White–Live cells

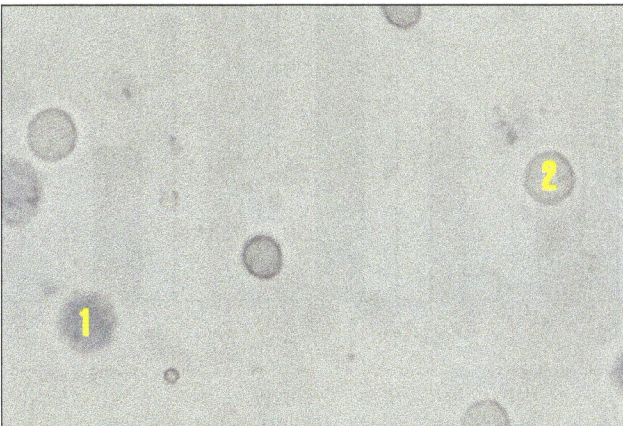

Cytotoxic nature of aqueous extract of *Sphaeranthus indicus* flower
(1) Blue–Dead cells; (2) White–Live cells

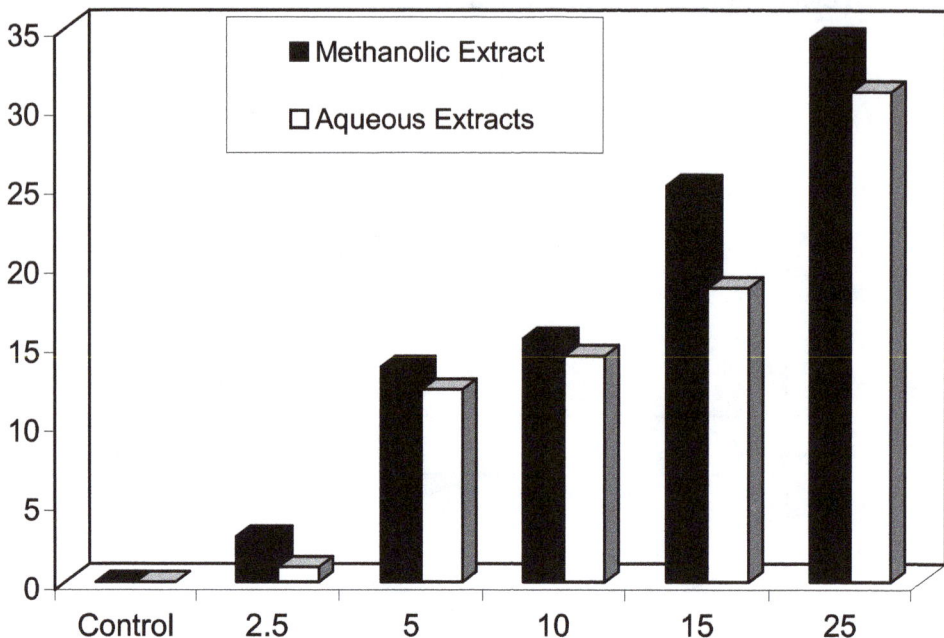

Figure 2.1L Cytotoxic effect of aqueous methanolic extract of *S. indicus* leaves on human liver cancer cell lines (HEPG2) by MTT method.

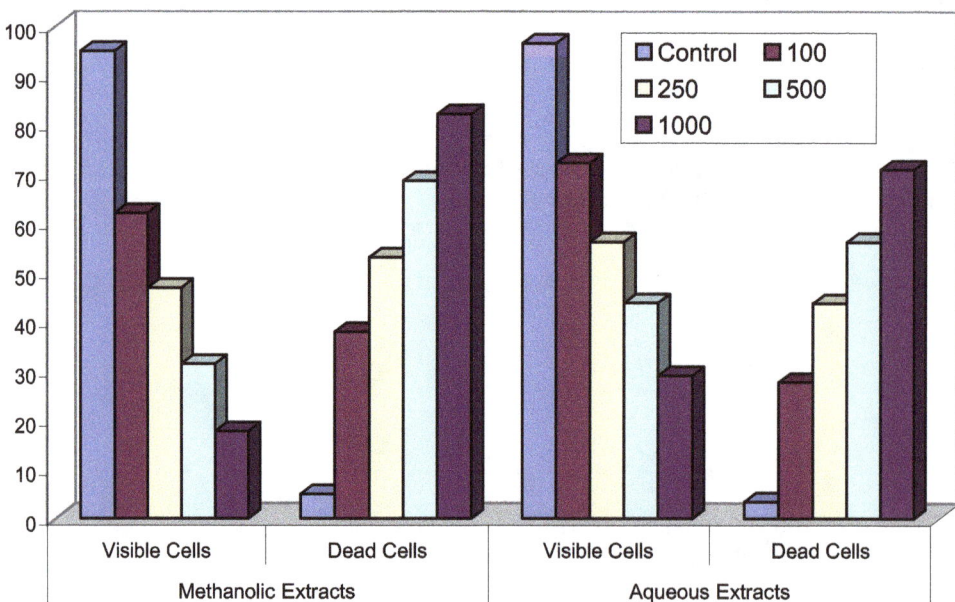

Figure 2.2: Cytotoxic effect of methanolic and aqueous extracts of *S. indicus* leaves on human liver cancer cell lines (HEPG2) by Tryphan blue dye exclusion method.

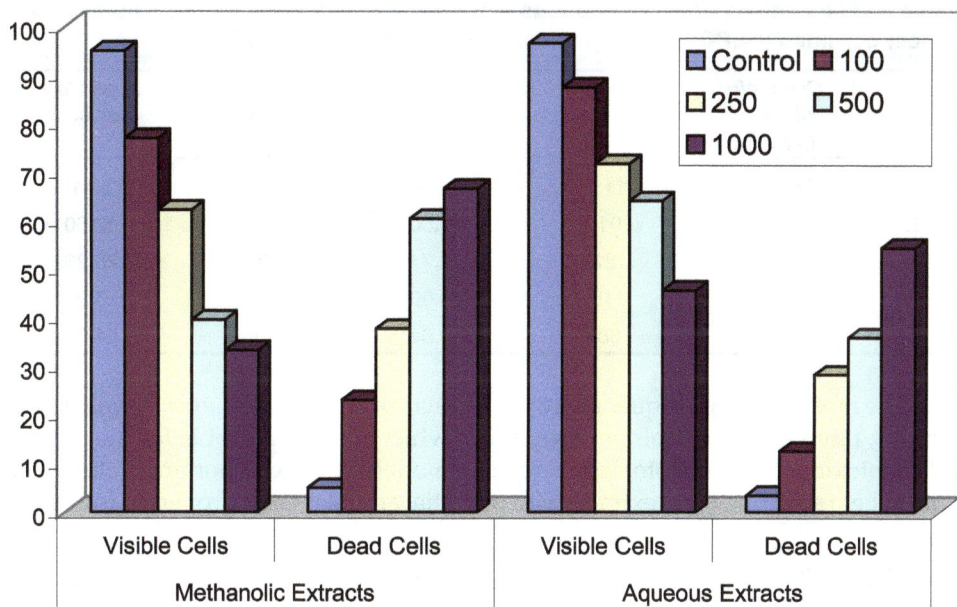

Figure 2.3: Cytotoxic effect of methanolic and aqueous extracts of *S. indicus* flowers on human liver cancer cell lines (HEPG2) by Tryphan blue dye exclusion method.

Table 2.11: Cytotoxic effect of methanolic extracts of *S. indicus* flowers on human liver cancer cell lines (HEPG2) by Tryphan blue dye exclusion method.

Sl.No.	Conc. of Plant Extract (μg/ml)	No. of Viable Cells	Viable Cells (per cent)	No. Dead Cells	Dead Cells (per cent)
1.	Control	114	95	6	5
2.	100	77	77	23	23
3.	250	61	62.25	37	37.75
4.	500	55	39.57	84	60.43
5.	1000	48	33.33	96	**66.67**

4. Discussion

Our phytochemical screening revealed that both aqueous and ethanolic extracts contained secondary metabolites such as alkaloids, flavanoids, saponins, tannins, steroids, reducing sugars and phenolic compounds. However, higher tannin content was found in the methanolic extract than aqueous extract (Thamilmaraiselvi *et al.*, 2011).

Table 2.12: Cytotoxic effect of aqueous extracts of *S. indicus* flowers on human liver cancer cell lines (HEPG2) by Tryphan blue dye exclusion method.

Sl.No.	Conc. of Plant Extract (µg/ml)	No. of Viable Cells	Viable Cells (per cent)	No. Dead Cells	Dead Cells (per cent)
1.	Control	113	96.59	4	3.41
2.	100	91	87.50	13	12.50
3.	250	122	71.77	48	28.23
4.	500	116	64.09	65	35.91
5.	1000	68	45.64	81	**54.36**

The methanolic and aqueous leaves extracts of *Sphaeranthus indicus* was rich in tannins, flavonoids, phenols, and saponins, which may responsible for this cancer cell cytotoxic activity. Cytotoxicity screening models provide important preliminary data to help selecting plant extracts with potential antineoplastic properties for future work.

The result of the present study suggests methanolic and aqueous extracts obtained from *Sphaeranthus indicus* leaves exhibit excellent cytotoxic activity on HepG2 cell line.

References

Ambavade, S.D., Mhetre, N.A., Tate, V.D. and Bodhankar, S.L., 2006. Pharmacological evaluation of the extracts of *Sphaeranthus indicus* flowers on anxiolytic activity in mice. *Indian J. Pharmacol.*, 38: 254–259.

Boham, B.A. and Kocipai-Abyazan, R., 1994. Flavonoids and condensed tannins from leaves of *Hawaiian vaccinium vaticulatum* and *V. calycinium*. *Pacific Sci.*, 48: 458–463.

Freshney, R.I., 1994. *Culture of Animal Cells: A Manual of Basic Technique*, 3rd edn. A John Wiley and Sons, Inc., Publication, New York, pp. 287–306.

Harborne, J.B., 1973. *Phytochemical Methods*. Chapman and Hall, Ltd., London, pp. 49–188.

Ikram, M., 1981. A review on the medicinal plants. *Hamdard*, 24: 102–129.

Jha, R.K., Garud, N. and Nema, R.K., 2009. Excision and incision wound healing activity of flower head alcoholic extract of *Sphaeranthus indicus* Linn. *in albino rats*. *Global J. Pharmacol.*, 3: 32–37.

Mosmann, T., 1983. Rapid colorimetric assay for cellular growth and survival: application to proliferation and cytotoxic assay. *J. Immuno. Met.*, 65: 55–63.

Ncube, N.S., Afolayan, A.J. and Okoh, A.I., 2008. Assessment techniques of antimicrobial properties of natural compounds of plant origin: Current methods and future trends. *Afr. J. Biotechnol.*, 7(12): 1797–1806.

Obdoni, B.O. and Ochuko, P.O., 2001. Phytochemical studies and comparative efficacy of the crude extracts of some Homostatic plants in Edo and Delta States of Nigeria. *Global J. Pure Appl. Sci.*, 8b: 203–208.

Siddiqui, A.A. and Ali, M., 1997. *Practical Pharmaceutical Chemistry*. CDS Publishers and Distributors, New Delhi, pp. 126–131.

Soffwara, E.A., 1994. *Medicinal Plant and Traditional Medicine in Africa*. University of IFE Press, Nigeria, pp. 1–23.

Thamilmaraiselvi, B., Ahamed John, S. and Theivandran, G., 2011. Preliminary phytochemical investigation and antimicrobial activity of *Sphaeranthus indicus* Linn. *Ad. Plant. Sci.*, 24(1): 81–85.

Thamilmaraiselvi, B., Ahamed John, S. and Theivandran, G., 2011. Preliminary phytochemical investigation and antimicrobial activity of *Clerodendrum phlomidis*. *Ad. Plant Sci.* (In Press).

Van-Burden, T.P. and Robinson, W.C., 1981. Formation of complexes between protein and Tannin acid. *J. Agric. Food Chem.*, 1: 77.

Scientific Basis of Herbal Medicine (2013)   *Pages 31–36*
Editor: Dr. Parimelazhagan Thangaraj
Published by: DAYA PUBLISHING HOUSE, NEW DELHI

Chapter 3

Evaluation of Antibacterial Activity of Leaf Extracts of the Medicinal Plant Species, *Hypochaeris radicata* L. (Asteraceae)

Jamuna Senguttuvan and Subramanium Paulsamy

*Department of Botany, Kongunadu Arts and Science College,
Coimbatore – 641 029, Tamil Nadu, India*

1. Introduction

Wound infections are common in developing countries due to poor hygienic conditions. *Staphylococcus aureus, Streptococcus pyogenes, Escherichia coli, Pseudomonas aeruginosa, Streptococcus faecalis* and *Klebsiella pneumoniae* are some important bacterial strains causing wound infections (Mertz and Ovington, 1993). A wide range of antibiotics are being used at present for treating wound infections (Lullmann *et al.*, 2000; Anonymous, 2004). Therefore, there is a continuous and urgent need to discover new antimicrobial compounds with diverse chemical structures and novel mechanisms of action (Abi Beaulah *et al.*, 2011). Medicinal plants represent a rich source of many potent and powerful drugs with wound healing properties (Emori and Gaynes, 1993; Srivastava *et al.*, 1996; Uniyal *et al.*, 2006). The screening of plant extracts has been of great interest to scientists in the search of new drugs for effective treatment of several infectious diseases (Dimayuga and Garcia, 1991). Therefore, plant extracts and phytochemicals with known antimicrobial properties can be of great significance in therapeutic treatments (Diallo *et al.*, 1999; Erdogrul, 2002; Rojas *et al.*, 2006).

The plant species, *Hypochaeris radicata* L. (Asteraceae) is commonly known as Catsear. It is native of Europe, distributed in high hills of Nilgiris, the Western Ghats, India (above 2000 m MSL). It has been extensively used in traditional medical practice as anticancer, anti-inflammatory, anti-diuretic and hepatoprotective activity and to treat kidney problems also. Moreover, it possesses wound healing activity and also used to cure various skin diseases caused by pathogens. The leaf and root part of this species are reported to have good antioxidant property also (Jamuna *et al.*, 2012). The aim of the present investigation was to evaluate the antibacterial activity of leaf extracts of *H. radicata* against some pathogenic bacterial strains which include both Gram-positive and Gram-negative types.

2. Materials and Methods

2.1. Plant Collection and Identification

The leaves of *H. radicata* were collected from Kattabettu, Nilgiris, the Western Ghats, India (2000 m above MSL). The authenticity of the plant was confirmed in Botanical Survey of India, TNAU Campus, Coimbatore by referring the deposited specimen. The voucher number is BSI/SRC/5/23/2010-11/Tech. 153.

2.2. Preparation of Plant Extracts

The dust free leaves of *H. radicata* were shade dried and powdered. About 50g of coarsely powdered plant materials (50 g/250 ml) were extracted in a soxhlet apparatus for 8 to 10 hours, sequentially with petroleum ether, chloroform, ethyl acetate, methanol and water separately in order to extract non-polar and polar compounds (Elgorashi and Van Staden, 2004). The extracts obtained were then concentrated and finally dried to a constant weight. Dried extracts were kept at 20°C until further use.

2.3. Preparation of Inoculum

Stock cultures were maintained at 4°C on slopes of nutrient agar. Active cultures for experiments were prepared by transferring a loop full of cells from the stock cultures to test tubes of nutrient agar medium and were incubated without agitation for 24 hrs at 37°C. The cultures were diluted with fresh nutrient agar broth to achieve optical densities corresponding to 2-10^6 colony forming units (CFU/ml) for bacteria.

2.4. Microbial Strains Used

In vitro antibacterial activity was examined for the leaf extract of the species, *H. radicata* against certain pathogenic bacteria which include the Gram-positive strains *viz.*, *Streptococcus faecalis*, *S. pyogenes* and *Enterococcus faecalis* and Gram-negative strains *viz.*, *Seratia marcescens*, *Klebsiella pneumoniae*, *Proteus vulgaris*, *P. mirabilis* and *Salmonella paratyphi*. All these bacteria were obtained from the Department of Biotechnology, Hindustan College of Arts and Science, Coimbatore.

2.5. Antibacterial Assay

The antibacterial activity of the leaf extracts was determined by using agar well diffusion method (Cruickshank *et al.*, 1975). The autoclaved media was poured in the sterilized petri plates. These plates were dried for a period of 20 minutes under aseptic condition before its use. Freshly grown cultures of the tested bacterial strains were streaked over the plates using a platinum wire inoculation loop. On sterile media plates, well of 5.0 mm diameter were punched with the help of a sterile cork borer. The extracts (50 µg/ml) were added into the wells by using micropipettes. Standard antibiotics, Amphicillin (50 µg/ml) was tested against the pathogens. The plates were incubated at 37°C for 24 hrs. After the incubation period, the diameter of the inhibition zones of each well was measured in millimeter.

Table 3.1: Antibacterial activity of leaf extracts of *Hypochaeris radicata* on certain human pathogenic bacteria.

Plant Extracts	Diameter of the Inhibition Zone (mm)							
	Gram-Positive			Gram-Negative				
	Streptococcus faecalis	S. pyogenes	Enterococcus faecalis	Seratia marcescens	Klebsiella pneumoniae	Proteus vulgaris	P. mirabilis	Salmonella paratyphi
Control*	27 ± 1^a	16 ± 1^a	23.33 ± 1.53^a	20.33 ± 0.58^a	23 ± 2^a	21.33 ± 1.53^a	19 ± 1^a	22.67 ± 2.52^a
Petroleum ether	9.67 ± 2.52^b	10 ± 1^b	7.67 ± 2.52^b	9.33 ± 2.08^b	9 ± 1^b	8.67 ± 1.53^b	7 ± 1^b	6 ± 1^b
Chloroform	20 ± 5^c	15.33 ± 3.06^a	12 ± 3^c	15.33 ± 3.06^c	12.67 ± 2.08^c	15 ± 1^c	10 ± 1^c	21.33 ± 3.21^a
Ethyl acetate	22 ± 5.57^{cd}	20.67 ± 0.58^c	18 ± 2.65^d	14 ± 1^c	20.33 ± 2.52^{cd}	16 ± 1^c	21 ± 1^{ad}	26 ± 4.58^c
Methanol	12 ± 1^{bc}	10 ± 1^b	12.67 ± 2.52^c	8 ± 1^b	13 ± 1.2^c	8.67 ± 1.53^b	9 ± 1^c	8.33 ± 3.06^d
Water	–	–	5.33 ± 0.58^e	–	–	9 ± 1^b	–	7.67 ± 2.52^{bd}

*Amphicillin, '–' indicates no activity.

Values were performed in triplicates and represented as mean±SD.

Mean values followed by different superscript in a column are significantly different (p<0.05).

2.6. Statistical Analysis

The antibacterial activity of *H. radicata* leaf extracts was indicated by clear zones of growth inhibition. All experiments were performed in triplicate and the results are presented as mean±SD (Standard Deviation). The significancy in the difference of mean was determined according to New Duncan's Multiple Range Test (Gomez and Gomez, 1976).

3. Results and Discussion

The antibacterial activity of leaf extracts of the study species, *H. radicata* against which human pathogenic bacteria which include both Gram-positive and Gram-negative bacteria are presented in Table 3.1. Among the five extracts tested, ethyl acetate extract had greater antibacterial potential, followed by chloroform, petroleum ether, methanol and water extracts. The highest zones of inhibition were observed for ethyl acetate and chloroform extracts against the Gram-positive bacterium *Streptococcus faecalis* (22±5.57) and (20±5) and Gram-negative bacterium *Salmonella paratyphi* (26±4.58) and (21.33±3.21) respectively. The petroleum ether and methanol extracts showed moderate activity for all bacterial strains tested which was ranging between 6mm and 10mm and 8mm and 13mm. In water extract the Gram-positive bacteria viz., *Streptococcus faecalis* and *S. pyogenes* and Gram-negative bacteria viz., *Seratia marcescens*, *Klebsiella pneumoniae* and *Proteus mirabilis* showed no inhibition. It was further observed that the inhibitory activity of ethyl acetate extract against the Gram-positive bacterium *Streptococcus pyogenes* and Gram-negative bacteria, *P. mirabilis* and *Salmonella paratyphi* was significantly greater than that of the standard drug, amphicillin.

From the above results it is known that the antibacterial activity of the ethyl acetate extract of *H. radicata* leaves is greater than that of the other solvent extracts, against both the Gram-positive and Gram-negative organisms. The higher antibacterial activity of the ethyl acetate and chloroform extracts may be due to the greater solubility of the phytochemicals in these organic solvents (De Boer *et al.*, 2005). The water extract of *H. radicata* leaves was practically ineffective against the test organisms. Several workers have reported that water extracts do not have much activity against bacteria (Martin, 1995; Paz *et al.*, 1995; Vlietinck *et al.*, 1995). It has been already reported that *H. radicata* contained more variety of flavonoids and phenolic acids and more required bioactive compounds for antibacterial activity (Zidon *et al.*, 2005). Most prominent antibacterial activity for many Asteraceae members has been well documented already (Chethan *et al.*, 2012; Dhanapal, 2011; Joshi, 2011).

4. Conclusion

It is concluded that the plant extract inhibited bacterial growth but their effectiveness is varied according to the solvents used. The antibacterial activity has been attributed to the presence of some active constituents in the extracts. It is interesting to note that the ethyl acetate extract of leaf of *H. radicata* could be used against the colonial growth of *Salmonella*. As *Salmonella* spp. are frequent candidates causing food-borne illnesses in addition to the typhoid and paratyphoid infection, the therapeutic value of this plant could be an effective remedy by inhibiting the growth of these bacteria. This result may provide a basis for the isolation of compounds from this species. Further studies are needed to identify the pure compounds and establish the mechanism of action for antibacterial activity of the plant extract.

References

Abi Beaulah, G., Mohamed Sadiq, A. and Jaya Santhi, R., 2011. Antioxidant and antibacterial activity of *Achyranthes aspera:* An *in vitro* study. *Annals of Biological Research,* 2(5): 662–670.

Anonymous, 2004. *British National Formulary.* Publication of British Medical Association UK.

Chethan, J., Sampath Kumara, K.K., Shailasree Sekhar and Prakash, H.S., 2012. Antioxidant, antibacterial and DNA protecting activity of selected medicinally important Asteraceae plants. *International Journal of Pharmacy and Pharmaceutical Sciences,* 4(2): 257–261.

Cruickshank, R., Duguid, J.P., Marmion, B.P.and Swain, R.H.A., 1975. *Medical Microbiology. Vol. 2: The Practice of Medical Microbiology,* 12th edn. Edinburgh, Churchill Livingston.

De Boer, H.J., Kool, A., Broberg, A., Mziray, W.R., Hedberg, I. and Levenfors, J.J., 2005. Antifungal and antibacterial activity of some herbal remedies from Tanzania. *Journal of Ethnopharmacol.,* 96(3): 461–469.

Diallo, D., Hveem, B., Mahmoud, M.A., Berge, G., Paulsen, B.S. and Maiga, A., 1999. An ethnobotanical survey of herbal drugs of Gourma district. *Mali Pharmaceutical Biology,* 37: 80–91.

Dimayuga, R.E. and Garcia, S.K., 1991. Antimicrobial screening of medicinal plants from Baja California Sur. Mexico. *Journal of Ethnopharmacology,* 31: 181–192.

Elgorashi, E.E. and Van Staden, J., 2004. Pharmacological screening of six Amaryllidaceae species. *Journal of Ethnopharmacology,* 90: 27–32.

Emori, T.G. and Gaynes, R.P., 1993. An overview of nosocomial infections, including the role of the microbiology laboratory. *Clinical Microbial Review,* 6: 428–442.

Erdogrul, O.T., 2002. Antibacterial activities of some plant extracts used in folk medicine. *Pharmaceutical Biology,* 40: 269–273.

Gomez, K.A.and Gomez, KA., 1976. *Statistical Procedure for Agricultural Research with Emphasis of Rice.* Los Bans, Philippines International Rice Research Institute Publisher, Laguna, Philippines, p. 294.

Jamuna, S., Paulsamy, S. and Karthika, K., 2012. Screening of *in–vitro* antioxidant activity of methanolic leaf and root extracts of *Hypochaeris radicata* L. (Asteraceae). *Journal of Applied Pharmaceutical Science,* 2(7): 149–154.

Joshi, R.K., 2011. *In vitro* antimicrobial activity of the essential oil of *Anaphalis contorta* Hook f. *International Journal of Research in Pure and Applied Microbiology,* 1(2): 19–21.

Lullmann, H., Mohr, K., Ziegler, A. and Bieger, D., 2000. *Color Atlas of Pharmacology*, 2nd edn. Publication of Thieme Stuttgart, New York, p. 166.

Martin, G.J., 1995. *Ethnobotany: A Methods Manual*. Chapman and Hall, London, England.

Mertz, P. and Ovington, L.G., 1993. Wound healing microbiology. *Dermatol. Clin.*, 11: 739–747.

Paz, E.A., Cerdeiras, M.P., Fernanadez, J., Ferrreira, F., Moyna, P., Soubes, M., Vazquez, A., Vero, S. and Zunino, L., 1995. Screening of Uruguayan medicinal plants for antimicrobial activity. *Journal of Ethnopharmacol.*, 45: 67–70.

Ramaiyan Dhanapal, 2011. Anthelmintic and antimicrobial activities of *Mikania micrantha* HBK. *Indian Journal of Pharmaceutical Science and Research*, 1(1): 1–3.

Rojas, J.J., Ochoa, V.J., Ocampo, S.A. and Munoz, J.F., 2006. Screening for antibacterial activity of ten medicinal plants used in Colombian folkloric medicine: A possible alternative in the treatment of non-nosocomial infections. *BMC complementary and Alternative Medicine*, 6: 2.

Srivastava, J., Lambert, J. and Vietmeyer, N., 1996. *Medicinal Plants: An Expanding Role in Development*. World Bank Technical paper. No. 320.

Uniyal, S.K., Singh, K.N., Jamwal, P. and Lal, B., 2006. Traditional use of medicinal plants among the tribal communities of Chhota Bhangal, Western Himalayan. *Journal of Ethnobiology and Ethnomedicine*, 2: 1–14.

Vlietinck, A.J., Van Hoof, L., Totte, J., Lasure, A., Vanden Berghe, D., Rwangabo, R.C. and Mvukiyumwami, J., 1995. Screening of hundred Rwandese medicinal plants for antimicrobial and antiviral properties. *Journal of Ethnopharmacology*, 46: 31–47.

Zidon, C., Schubert., B. and Stuppner, H., 2005. Altitudinal differences in the contents of phenolics in flowering heads of three members of the tribe Lactuceae (Asteraceae) occurring as introduced species in New Zealand. Biochemical and Systematic Ecology 33, 855–872.

Scientific Basis of Herbal Medicine (2013) Pages 37–45
Editor: Dr. Parimelazhagan Thangaraj
Published by: DAYA PUBLISHING HOUSE, NEW DELHI

Chapter 4

Effect of *Punica granatum* on a Dandruff Causing Fungus: A Phytopathogenic and Few Human Pathogenic Bacterial Strains

M. Faizal Ahamed, K. Sundaravel,
R. Ramachandran and R. Lalitha

*Department of Biotechnology,
Achariya Arts and Science College, Villianur,
Puducherry – 605 110, India*

1. Introduction

Punica granatum L. (Pomegranate) belongs to the family *Punicaceae* which has native of Iran but cultivated throughout India. According to Indian Herbal System, all parts of pomegranate including roots, leaves, flowers, rind, seeds and the reddish brown bark are used medicinally. Fruit is used to treat stomachic, digestive and uterine disorders; bark of stem and root is anthelmintic; rind of fruit, bark of stem and root is antidiarrheal; powdered flower buds are used to treat bronchitis (C.P. Khare, 2007). The potential therapeutic properties of pomegranate are wide-ranging and include treatment and prevention of cancer, cardiovascular disease, diabetes, dental conditions, erectile dysfunction, and protection from ultraviolet (UV) radiation. Other potential applications include infant brain ischemia, Alzheimer's disease, male infertility, arthritis, and obesity (Julie Jurenka, 2008).

1.1. Human Bacterial Pathogens

Greater part of bacteria are harmless or beneficial, rather a few bacteria are pathogenic. One of the bacterial diseases with highest disease burden is tuberculosis, caused by the bacterium *Mycobacterium tuberculosis*, which kills about 2 million people a year. Consequent to tuberculosis, Pneumonia is also a globally important disease, which can be caused by bacteria such as *Streptococcus* and *Pseudomonas*. As a part of the present study, three human bacterial pathogens were chosen to examine their antibiotic sensitivity against a set of medicinal plants. *Pseudomonas aeruginosa* is a gram negative, aerobic, rod shaped bacteria which causes generalized inflammation and sepsis in human; *Klebsiella pneumonia* is a gram negative, facultative anaerobic rod shaped bacteria which is the causative organism of inflammation in human lungs, hemorrhage, thick, bloody, mucoid sputum; *Proteus bacilli* is a gram negative proteobacteria which is responsible for urinary and septic infections, often nosocomial.

Some bacterial diseases have been overpowered, while many new bacterial pathogens have been recognized in the past 30 years. At the same time, many previously existing bacterial pathogens have emerged with new forms of virulence and new patterns of resistance to antimicrobial agents. Henceforth, novel approaches of research and study are needed to control both old and new bacterial pathogens.

1.2. Citrus Canker

Citrus canker is a disease mainly caused by the phytopathogen, *Xanthomonas*, consists of three species. Among the three species, as a causative of citrus canker, *Xanthomonas citri* subsp. *citri* has a wide range of host specificity. It significantly affects the vitality of citrus trees, causing leaves and fruit to drop prematurely, however not harmful to human. The severity of this infection varies with different species and varieties and the prevailing climatic conditions. Citrus canker is endemic in India, Japan and other South- East Asian countries, from where it has spread to all other citrus producing continents except Europe (Das, 2003).

Owing to the canker lesions on the surface of the fruits, they become unacceptable for market. Worldwide, millions of dollars are spent annually on prevention, quarantines, eradication programs, and disease control. Integrated pest management for canker disease consists of (i) using canker-free nursery stock, (ii) Pruning all the infected twigs before monsoon and burning them, (iii) Periodical spraying of suitable copper-based bactericides (to reduce inoculum build-up on new flushes and to protect expanding fruit surfaces from infection) along with an insecticide (to control insect injury), (iv) taking some precautions to reduce the risk of spread of disease in orchards and nurseries and (v) by evolving canker-resistant varieties suited to local environmental conditions (Das and Singh, 1999, 2001). However, studies on biological control of citrus canker are still in a preliminary stage.

1.3. Fungal Pathogen

Malassezia furfur (*Pityrosporum ovale*), is a lipophilic, yeast-like fungus, occurring in human skin (opportunistic pathogen) and a major causative of dandruff. Dandruff is a minor infection, causes small white flakes of skin on the scalp that falls off in flakes. *M. furfur* needs lipid to grow, found abundantly in oily skin. Neem, black pepper and egg oil contain selenium sulphide, zinc pyrithione, ketoconazole which are traditionally used as a treatment for dandruff. The signs and symptoms include: white flakes on shoulders of dark clothing, itchy scalp, dry facial skin, recurrent ear "eczema", facial rash by eyebrows, nose, and ears. Oily scalp and facial skin with dry flakes, eyebrow dandruff, beard dandruff, chest rash with dry flakes and red spots. Dandruff control may require long-term skin and hair care for best results.

For the present study ten medicinal plants which include *Punica granatum* were identified, collected and crude extracts were obtained from different parts of them by methanol extraction method. Antibacterial, antifungal activities of them have been investigated and documented.

2. Materials and Methods

2.1. Collection of Plant Material

Fresh, disease free samples of ten medicinal plants such as *Ficus religiosa* (leaf), *Leucas aspera* (leaf and flower), *Solanum xanthocarpum* (leaf and fruit), *Piper betle* (leaf), *Phyllanthus niruri* (entire plant), *Passiflora incarnate* (leaf), *Punica granatum* (leaf and flower), *Brassica nigra* (seed), *Zingiber officinale* (rhizome), and *Solanum torvum* (leaf) were identified and collected from in and around Pondicherry. The different parts were washed thoroughly 2-3 times in running water and once with sterile distilled water. Then they were air-dried on sterile blotter under shade. The dried samples were powdered and stored in clean containers.

2.2. Solvent Extraction

Crude extracts were obtained by methanol solvent extraction method (1 g powder: 4ml solvent). The samples were kept in rotary shaker for five days at 37°C and bioactive compounds were collected by evaporating the solvent using rotary evaporator.

2.3. Determination of Antibacterial Activity against Human Pathogen

Three human bacterial pathogenic samples, *Pseudomonas aeruginosa*, *Klebsiella pneumonia* and *Proteus bacilli*, were obtained from JIPMER, Pondicherry. The bacterial inoculum were pre-cultured in nutrient broth overnight in a rotary shaker at 37°C, centrifuged at 10,000 rpm/5 min., pellets were suspended in double distilled water and the cell density was standardized spectrophotometrically (A_{610}). The isolated bacteria were inoculated in Nutrient Agar medium (2.95 g/100 ml) by spread method and the sterile discs (6 mm in diameter) (HiMedia) infused with the extracts (disc diffusion method) were placed on test organisms-spread plates. After inoculation, antibacterial assay plates were incubated at 37°C/24 h. The diameter of the inhibition zones were measured in mm.

2.4. Determination of Antibacterial activity against Phytopathogen

The inoculum of *Xanthomonas citri* has been collected from infected leaves and fruits of lemon plant and subsequently cultured for ensuring their unambiguousness. The isolated bacterium was inoculated in Nutrient Agar medium (2.95 g/100 ml) by spread method and the sterile discs (6 mm in diameter) (HiMedia) infused with the extracts were placed on test organisms-spread plates by disc diffusion method. After inoculation, antibacterial assay plates were incubated at 37°C for 24 h. The diameter of the inhibition zones were measured in mm.

2.5. Antimycotic Activity against Fungal Pathogen

Dandruff from the infected person was collected and cultured in Sabouraud's dextrose medium (6g/100 ml) mixed with butter (2 g/100 ml). The isolated *Malassezia furfur* was inoculated in the same set of medium by spread method and the sterile discs (6 mm in diameter) (HiMedia) infused with the extracts were placed on test organisms-spread plates. After inoculation, antimycotic assay plates were incubated at 37°C/24 h. The diameter of the inhibition zones were measured in mm.

3. Results and Discussion

Among ten medicinal plants utilized for the present study to investigate their antimicrobial properties, six of them, *Leucas aspera* (leaf), *Piper betle* (leaf), *Phyllanthus niruri* (whole plant), *Punica granatum* (leaf and flower), *Brassica nigra* (seed), *Solanum torvum* (leaf) (Table 4.1) had the antimicrobial potentials against different bacterial and fungal pathogens. However, remarkable antimicrobial potentials have been recognized in *Punica granatum* leaf and its flower extracts in opposition to all the pathogens (Figure 4.1).

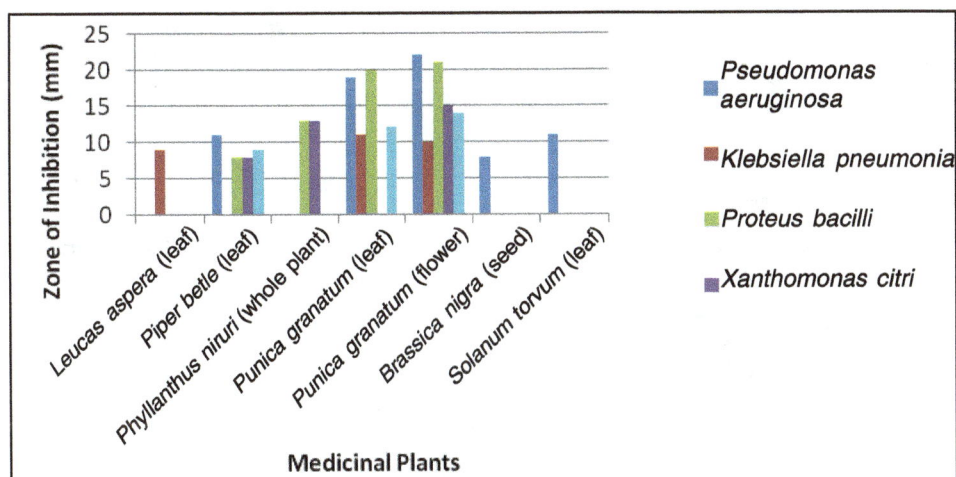

Figure 4.1: Antimicrobial activity against bacterial and fungal pathogens.

3.1. Antibacterial Activity against Human Bacterial Pathogens Tested by Disc Diffusion Assay

Results obtained from the present study revealed that six medicinal plants extracts possess potential antibacterial activity against human pathogens, *Pseudomonas aeruginosa*, *Klebsiella pneumonia* and *Proteus bacilli* (Table 4.1). Among the six plants, *Punica granatum* (both leaf and flower extracts) exhibited highest inhibitory activity against *Pseudomonas aeruginosa* (19 and 22 mm), *Klebsiella pneumonia* (11 and 10 mm), and *Proteus bacilli* (20 and 21 mm) (Plates 4.1 to 4.3). Moderate inhibitory activity has been noticed in *Piper betle* (11 mm), *Solanum torvum* (11 mm), *Brassica nigra* (8 mm) extracts against *Pseudomonas aeruginosa*, *Leucas aspera* leaf extract (9 mm) against *Klebsiella pneumonia*, and *Phyllanthus niruri* (13 mm), Piper betle (8 mm) extracts against *Proteus bacilli* (Table 4.1).

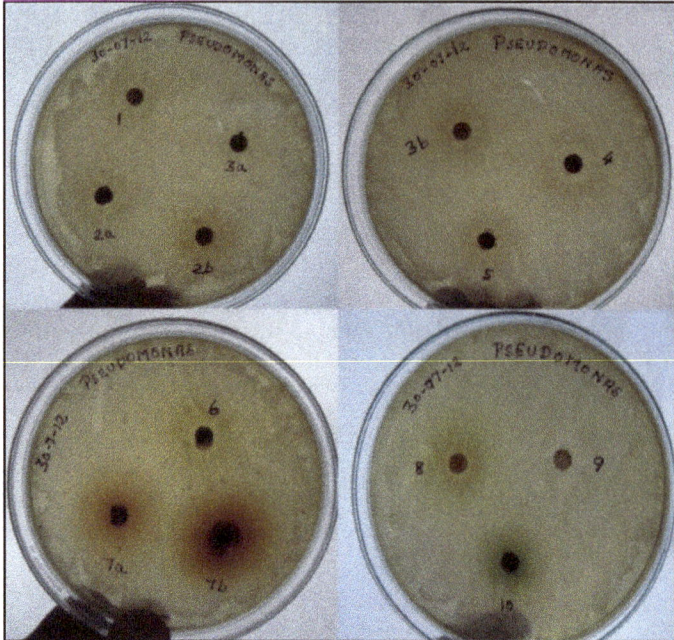

Plate 4.1: Antibacterial activity against human bacterial pathogen, *Pseudomonas aeruginosa* tested by disc diffusion assay (4, 7a, 7b, 8 and 10).

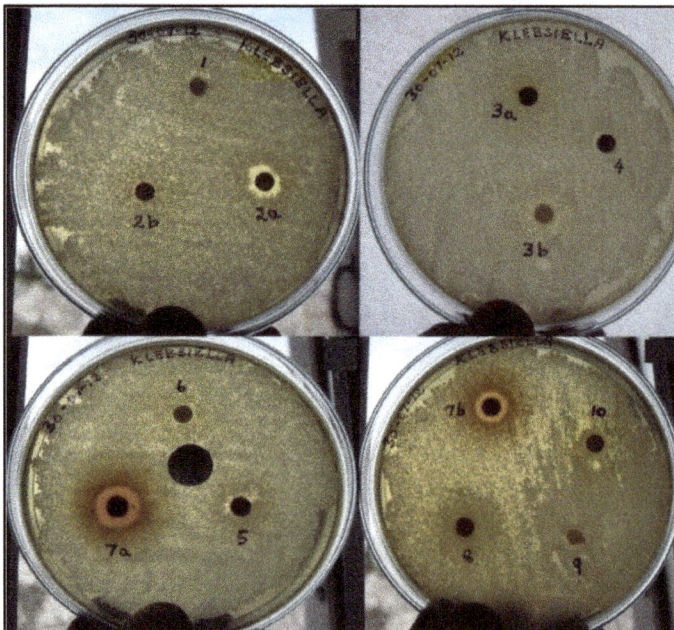

Plate 4.2: Antibacterial activity against human bacterial pathogen, *Klebsiella pneumonia* tested by disc diffusion assay (2a, 7a and 7b).

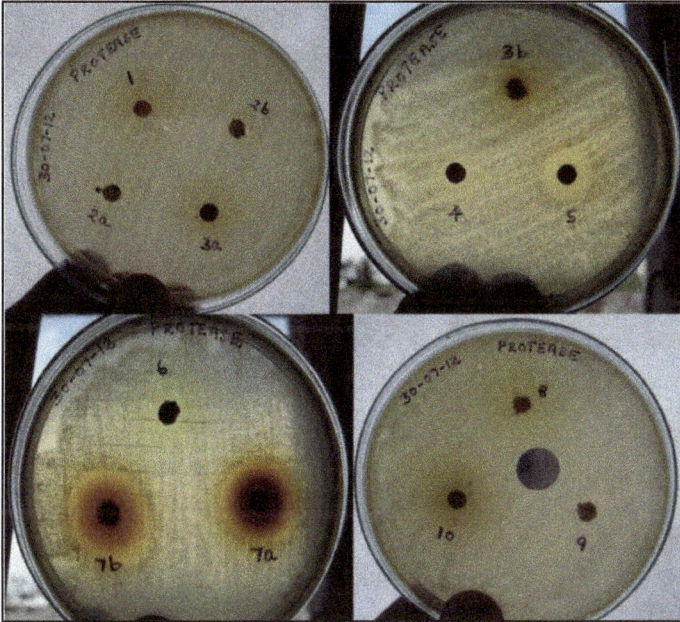

Plate 4.3: Antibacterial activity against human bacterial pathogen, *Proteus bacilli* tested by disc diffusion assay (4, 5, 7a and 7b).

Plate 4.4: Antibacterial activity against phytopathogen, *Xanthomonas citri* tested by disc diffusion assay (4, 5 and 7b).

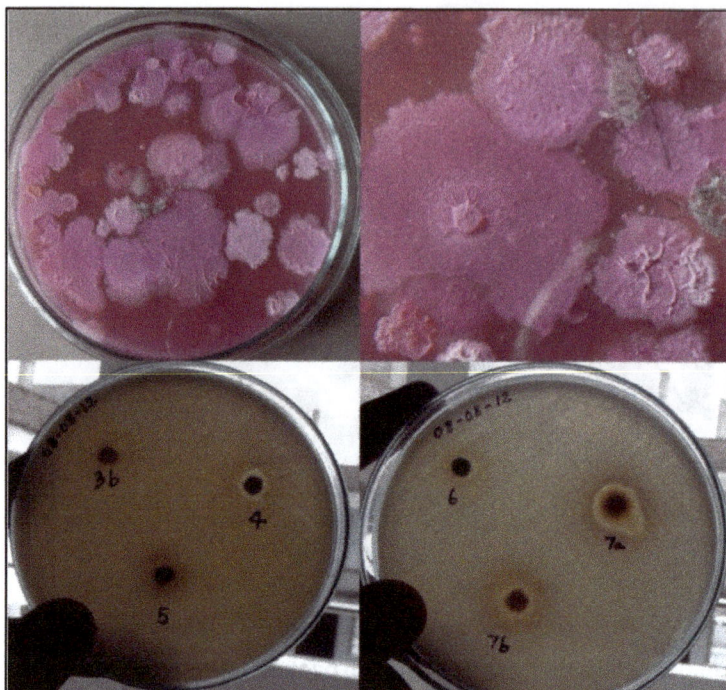

Plate 4.5: Antimycotic activity against fungal pathogen, *Malassezia furfur* tested by disc diffusion assay (4, 7a and 7b).

Table 4.1: Antimicrobial activity against human pathogen, phytopathogen and fungal pathogen tested by disc diffusion assay.

Sl.No.	Name of the Medicinal Plant	Parts Used	Zone of Inhibition (mm)				
			A*	B*	C*	D*	E*
1.	Ficus religiosa	Leaf	—	—	—	—	—
2a.	Leucas aspera	Leaf	—	09	—	—	—
2b.	Leucas aspera	Flower	—	—	—	—	—
3a.	Solanum xanthocarpum	Leaf	—	—	—	—	—
3b.	Solanum xanthocarpum	Fruit	—	—	—	—	—
4.	Piper betle	Leaf	11	—	08	08	09
5.	Phyllanthus niruri	Whole plant	—	—	13	13	—
6.	Passiflora incarnate	Leaf	—	—	—	—	—
7a.	Punica granatum	Leaf	19	11	20	—	12
7b.	Punica granatum	Flower	22	10	21	15	14
8.	Brassica nigra	Seed	08	—	—	—	—
9.	Zingiber officinale	Rhizome	—	—	—	—	—
10.	Solanum torvum	Leaf	11	—	—	—	—

*: A: *Pseudomonas aeruginosa*; B: *Klebsiella pneumonia*; C: *Proteus bacilli*; D: *Xanthomonas citri*; E: *Malassezia furfur*.

3.2. Antibacterial Activity against Phytopathogen, *Xanthomonas citri* Tested by Disc Diffusion Assay

Observations revealed that *Punica granatum* flower extract (15 mm) exhibited highest inhibitory activity, while *Phyllanthus amarus* (13 mm), *Piper betle* extracts (8 mm) showed moderate inhibitory activity against *Xanthomonas citri* (Table 4.1) (Plate 4.4).

3.3. Antifungal Activity against Fungal Pathogen, *Malassezia furfur* Tested by Disc Diffusion Assay

From the examination it has been noted that *Punica granatum* (both leaf and flower extracts) exhibited highest inhibitory activity against *M. furfur* (12 and 14 mm) while moderate inhibitory activity noted in *Piper betle* (9 mm) (Table 4.1) (Plate 4.5).

4. Conclusion

Phytomedicine always play a vital role in developing novel chemotherapeutic agents. The primary pace towards this aspiration is *in vitro* antimicrobial assay. Many reports are available on the antiviral, antibacterial, antifungal, anthelmintic, antimolluscal and anti-inflammatory properties of plant medicines (Palombo and Semple, 2001; Mahesh and Satish, 2008). Pomegranate has antimicrobial, antioxidant, anticarcinogenic and anti-inflammatory properties, which suggest its possible use as a therapy for prevention and treatment of several types of cancer and cardiovascular disease (Julie Jurenka, 2008). In addition to its ancient historical uses, pomegranate is used in several systems of medicine for a variety of ailments. Previous reports stated that Pomegranate flower possess the active compounds such as gallic acid, ursolic acid (Huang *et al.*, 2005), triterpenoids, including maslinic and asiatic acid (Batt and Rangaswami, 1973) and other unidentified constituents as well. Furthermore Pomegranate leaves possess tannins (punicalin and punicafolin) and flavone glycosides, luteolin and apigenin (Nawwar *et al.*, 1994).

Generally the methanol extract of medicinal plants have the highest activity against both bacterial and fungal isolates (Igbinosa *et al.*, 2009). From the present study as well it has been proved that methanol extraction of *Punica granatum* has exhibited maximum inhibitory activity against different bacterial and fungal pathogens. Numerous therapeutic uses of *Punica granatum* has led to a variety of clinical trials over the last decade. Contemporary study also suggests that the active compound of *Punica granatum* can be isolated and utilized in commercial shampoos as an anti-dandruff component owing to its antifungal property against the fungal pathogen which is the causative of dandruff infection, *Malassezia furfur*.

In addition, the antibacterial activity of *Punica granatum* against a phytopathogen, *Xanthomonas citri*, suggested that intercropping of it along with crop plants may enhance the antibacterial activity of crop plants. Since few reports are only available about the role of phytomedicine in crop protection, present investigation about *Punica granatum* set alight on a novel manipulation of it in the field of agriculture for crop improvement and for crop protection. On the other hand, the antibacterial activities of *Punica granatum* against human pathogens have been proved since long back by a variety of persistent researches. In a favourable manner, constructive results were obtained from crude extract of *Punica granatum* leaf and its flower in opposition to three human bacterial pathogens such as *Pseudomonas aeruginosa, Klebsiella pneumonia* and *Proteus bacilli*. It recommends day to day consumption of *Punica granatum* as a food additive remarkably improves resistance in human against pathogens. However, further investigation is required to identify the active compound of *Punica granatum,* which possess all these antimicrobial activities.

References

Batt, A.K. and Rangaswami, S., 1973. Crystalline chemical components of some vegetable drugs. *Phytochemistry,* 12: 214.

Das, A.K., 2003. Citrus canker: A review. *J. Appl. Hort.,* 5(1): 52–60.

Das, A. K. and Singh, Shyam, 1999. Management of bacterial canker in acid lime. *Intensive Agriculture,* 36(11–12): 28–29.

Das, A.K. and Singh, Shyam, 2001. Managing citrus bacterial diseases in the state of Maharashtra. *Indian Hort.,* 46(2): 11–13.

Huang, T.H., Yang, Q. and Harada, M., 2005. Pomegranate flower extract diminishes cardiac fibrosis in Zucker diabetic fatty rats: Modulation of cardiac endothelin–1 and nuclear factor–kappaB pathways. *J. Cardiovasc. Pharmacol.,* 46: 856–862.

Igbinosa, O.O., Igbinosa, E.O. and Aiyegoro, O.A., 2009. Antimicrobial activity phytochemical screening of stem bark extracts from *Jatropha curcas* (Linn). *African Journal of Pharmacy and Pharmacology,* 3(2): 58–62.

Julie Jurenka, M.T., 2008. Therapeutic applications of pomegranate (*Punica granatum* L.): A review. *Alternative Medicine Review,* 13(2).

Khare, C.P., 2007. *Indian Medicinal Plants: An Illustrated Dictionary.* Springer.

Mahesh, B. and Satish, S., 2008. Antimicrobial activity of some important medicinal plant against plant and human pathogens. *World Journal of Agricultural Sciences,* 4: 839–843.

Nawwar, M.A., Hussein, S.A. and Merfort, I., 1994. NMR spectral analysis of polyphenols from *Punica granatum. Phytochemistry,* 36: 793–798.

Palombo, E.A. and Semple, S.J., 2001. Antibacterial activity of traditional medicinal plants. *Journal of Ethnopharmacology,* 77: 151–157.

Scientific Basis of Herbal Medicine (2013)
Editor: Dr. Parimelazhagan Thangaraj
Published by: DAYA PUBLISHING HOUSE, NEW DELHI

Pages 47–59

Chapter 5

Phytochemical and Antimicrobial Activities of *Gnaphalium polycaulon*

K. Shanmugapriya and T. Thayumanavan

*School of Biotechnology,
Dr. G.R. Damodaran College of Science,
Coimbatore – 641 014, Tamil Nadu, India*

1. Introduction

Aromatic and medicinal plants are sources of diverse nutrient and non-nutrient molecules, many of which display antioxidant and antimicrobial properties that can protect the human body against both cellular oxidation reactions and pathogens (Cowan, 1999). The use of plants for medicinal purposes can be traced back to earlier civilization (Le Strange *et al.*, 1977). Medicinal plants have a global distribution although they are most abundant in the tropics (Calixto *et al.*, 2000). In as much as the period where medicinal plants were being used has been long, the infectious diseases are as old as the medicinal plants (Azoro, 2002). The primary benefit of the utilization of plant derived medicine is that they are relatively safer than synthetic alternatives, offering very good therapeutic treatment (Robbers *et al.*, 1996). Flavonoids are capable of treating certain physiological disorder and diseases (Okwu, 2004).

Microorganisms are closely associated with the health and welfare of human beings, some microorganisms are beneficial and others are detrimental (Hufford *et al.*, 1993). Throughout the history of mankind, many infectious diseases have been known to be treated with herbal remedies. This results to a never ending and urgent need to discover new antimicrobial compounds with different chemical structure and new mechanisms of action for re-emerging and new infectious diseases (Rojas *et al.*, 2003). Therefore, researchers are increasingly turning their keen attention towards folk medicine from plants which leads into developing better natural drugs against microbial infections (Benkeblia *et al.*, 2004). Several medicinal plants are being screened for their potential microbial activity based on increasing failure of chemotherapy and antibiotic resistance exhibited by pathogenic agents (Colombo *et al.*, 1996). According to World Health Organisation, 65-80 per cent of the world populations rely on traditional medicine to treat various diseases. To date, many plants have been claimed to pose beneficial health effects such as antioxidant and antimicrobial properties. With the emergence of multiple strains of antibiotic resistance microorganism, great interest has been generated in search for potential compounds from plants for therapeutic, medicinal, aromatic and aesthetic uses (Ashebir and Ashenati, 1999).

Thin layer chromatography (TLC) is a chromatography technique used to separate mixtures. There are several essential features of TLC, connected to its simplicity that is important in the analysis of pharmaceutical preparations. It focuses mainly on steroid analysis in environmental materials such as pharmaceuticals, plant products and other biological specimens and employed for the routine analysis of steroids in pharmaceutical formulations and in bulk drug preparations as well as for the quality assurance of related extracts and market samples (Santoyo *et al.*, 2006). Phytochemicals are natural and non-nutritive bioactive compounds produced by plants that act as protective agents against external stress and pathogenic attack (Sofowora, 1993). The plant-derived phytochemicals with therapeutic properties could be used as single therapeutic agent or as combined formulations in drug development (Yen *et al.*, 1995).

Gnaphalium polycaulon is a genus of flowering plants in the asterceae family of compositae type, worldwide distribution and is mostly found in temperate regions, although some are found on tropical mountains or in the subtropical regions of the world. *G.polycaulon* was described in 1807 by Christiaan Hendrik Persoon. The entire plant is harvested during flowering and is used to make herbal and homeopathic remedies (Adhikari and Babu, 2008). Practitioners prescribe the herb for respiratory, digestive, and musculoskeletal conditions as well as an aid to quit smoking. The homeopathic remedy has no known side effects; when taken as a tea. Species in this genus are said to have anti-inflammatory, astringent, and antiseptic properties and are often prescribed as an herbal supplement for colds, flu, pneumonia, tonsillitis, larygitis, and congestion (Bhagwati uniyal and Vandana shiva, 2005). It is a popular treatment for respiratory problems and neuritis among the Lumbee Native American tribe. It may be used in place of tobacco cigarettes. Patients with rheumatism, diarrhoea and an increase in urination, combined with sporadic upper jaw pain, may benefit from *Gnaphalium polycaulon plant* (Sandeep Acharya, 2011).

Biodiversity studies reveal that the plant kingdom has not been exhausted based on the species of medicinal plants which are yet to be discovered. The investigations of biological activity and chemical composition of medicinal plants as a potential source of natural antioxidants are numerous. So, this medicinal plant was chosen for our present study with main objectives to screen the phytochemicals constituents and to separate the components present in the plant by Thin Layer Chromatography and to evaluate the antimicrobial study.

2. Materials and Methods

The methodology adopted for the study entitled "Preliminary evaluation of phytochemical constituents and antimicrobial activity of *G. polycaulon* is given below:

2.1. Chemicals Required

All chemicals used for this study were high quality analytical grade reagents. The solvents such as ethanol, methanol, petroleum ether and hexane were purchased from S.D. Fine Chemicals Pvt. Ltd., Sigma Chemicals, Lobe chemicals, Merck Chemical Supplies, Nice Chemicals and HiMedia. All other chemicals used for the study were obtained commercially and were of analytical grade.

2.2. Collection of Plant Material

The fresh leaves, stem and flower of *G.polycaulon* plant were collected from Kodanadu near Kotagiri in The Nilgiri district.

2.3. Preparation of Extract

Plant materials were washed, air dried and coarsely powdered. Forty grams of the powdered sample was extracted sequentially for 3 h into 250 ml of methanol, ethanol and water. Resulting extracts were evaporated and concentrated to dryness using the rotary evaporator. Powder was weighed and dissolved in the solvents used for extraction: methanol, ethanol and water separately and stored at 4°C.

2.4. Qualitative Analysis of Phytochemical

The extracts were used for preliminary screening of phytochemicals such as carbohydrates (Molisch's test and Benedict's test), protein (Biuret test), aminoacids (Ninhydrin test), alkaloids (Wagner and Dragendorff's tests), flavonoids, tannins, saponins, cardiac glycosides (Keller-Killiani test), terpenoids (Salkowski test), steroids (Liberman-Burchard Test), gum (Molisch's test), anthraquinone (Borntrager's test) and phlobatinin by standard methods (Harborne *et al.*, 1973; Odebiyi and Ramstard, 1978).

2.5. Quantitative Analysis of Phytochemicals

2.5.1. Determination of Total Phenolic Content

The total phenolic content of all leaf extracts of the plant material was estimated according to the method of Siddhuraju and Becker method (2003). Different aliquots of the extracts were taken in a test tube and made up to the volume of 1 ml with distilled water. Then 0.5 ml of Folin-Ciocalteu phenol reagent (1: 1 with water) and 2.5 ml of sodium carbonate solution (20 per cent) were added sequentially in each tube in triplicate manner. Soon after vortexing the reaction mixture, the test tubes were placed in dark for 40 min and the absorbance was recorded at 725 nm against a reagent blank. Total phenolic contents were determined as a Gallic acid equivalent (GAE) based on Folin-Ciocalteau calibration curve using Gallic acid (ranging from 50 to 1000 mg/ml) as the standard and expressed as mg Gallic acid equivalent per gram. All tests were carried out in triplicate.

2.5.2. Determination of Total Flavonoid Content

The flavonoid content of the plant extract was determined by a colorimetric method (Xia Liu *et al.*, 2005 and Zhishen *et al.*, 1999) with minor modification. Different concentration of the plant extract was prepared by diluting the stock solution

(4000 µg/ml) with deionized water. Each sample (100 µl) was diluted with distilled water (200 µl). Sodium nitrite (5 per cent; 30 µl) was added to the samples and then at 5 minutes, aluminium chloride (10 per cent; 30 µl) and at 6 minutes, sodium hydroxide (1 M; 200 µl) were added to the mixture. Finally, 400 µl of deionized water was added. The absorbance was recorded at 510 nm. Quercetin was used as the standard to calculate the concentrations of flavonoid content and the values were expressed as mg Quercetin equivalents/g of extract. The analysis was performed in triplicates.

2.5.3. Determination of Chlorophyll

The chlorophyll content in the various plants parts was estimated by the method of Witham *et al.* (1971). Chlorophyll was extracted from 1 g of the sample using 20 ml of 80 per cent acetone. The supernatant was transferred to a volumetric flask after centrifugation at 5000 rpm for 5 minutes. The extraction was repeated until the residue became colourless. The volume in the flask was made up to 100 ml with 80 per cent acetone. The absorbance of the extract was read in a spectrophotometer at 645 and 663 nm against 80 per cent acetone blank.

The amount of total chlorophyll in the sample was calculated using the formula,

$$\text{Total chlorophyll} = 20.2\,(A_{645}) + 8.02\,(A_{663}) \times \frac{V}{1000 \times W}$$

where,

V = Final volume of the extract

W = Fresh weight of the leaves

The values are expressed as mg chlorophyll/g sample.

2.5.4. Determination of Alkaloid

The Alkaloid was determined by the methods of Harborne (1973). Five grams of the sample was weighed into a 250 ml beaker and 200 ml of 10 per cent acetic acid in ethanol was added and allowed to stand for 4 h. Then filtered and the extract was concentrated on a water bath to one-quarter of the original volume. Concentrated ammonium hydroxide was added drop wise to the extract until the precipitation was complete. The whole solution was allowed to settle and the precipitated was collected and washed with dilute ammonium hydroxide and then filtered. The residue is the alkaloid was dried and weighed. All tests were carried out in triplicate.

2.5.5. Determination of Tannin

The Tannin was determined by the methods of Van-Burden and Robinson (1981). Five hundred milligrams of the sample was weighed into a 50 ml plastic bottle. Fifity millilitres of distilled water was added and shaken for 1 h in a mechanical shaker. Then filtered into a 50 ml volumetric flask and made up to the mark. Then 5 ml of the filtered was pipetted out into a test tube and mixed with 2 ml of 0.1 M $FeCl_3$ in 0.I N HCl and 0.008 M potassium ferrocyanide. All tests were carried out in triplicate. The absorbance was measured at 120 nm within 10 min.

2.5.6. Determination of Saponin

The Saponin was determined by the methods of Obadoni and Ochuko (2001). Twenty grams of plant sample were put into a conical flask and 100 cm³ of 20 per cent aqueous ethanol were added. The samples were heated over a hot water bath for 4 h with continuous stirring at about 55°C. The mixture was filtered and the residue re-extracted with another 200 ml 20 per cent ethanol. The combined extracts were reduced to 40 ml over water bath at about 90°C. The concentrate was transferred into a 250 ml separatory funnel and 20 ml of diethyl ether was added and shaken vigorously. The aqueous layer was recovered while the ether layer was discarded. The purification process was repeated. 60 ml of n-butanol was added. The combined n-butanol extracts were washed twice with 10 ml of 5 per cent aqueous sodium chloride. The remaining solution was heated in a water bath. After evaporation, the samples were dried in the oven to a constant weight. The saponins content was calculated as percentage.

2.6. Thin Layer Chromatography

The powdered plant sample was lixiviated in ethanol on rotary shaker (180 thaws/min) for 24 h (Santoyo *et al.*, 2006). The condensed filtrate was used for chromatography. The phenols were separated using chloroform and methanol (27: 0: 3) solvent mixture. The alkaloids spots were separated using the solvent mixture chloroform and ethanol (15: 1). The flavonoid spots were separated using chloroform and ethanol (19: 1) solvent mixture. The saponins were separated using chloroform, glacial acetic acid, methanol, water (3: 1.5: 0.6: 0.2) solvent mixture. The glycosides were separated using chloroform, methanol, conc. ammonia (40: 10: 2) solvent mixture. The colour of spots was identified. The retardation factors (*Rf*) of all components are reported. *Rf* value is calculated by using the formula.

$$\text{Retardation factors } (Rf \text{ value}) = \frac{\text{Distance travelled by solute}}{\text{Distance travelled by solvent from the origin}}$$

2.7. Antimicrobial Activity

2.7.1. Test Organisms

The following cultures of Gram negative bacteria (*Salmonella typhimurium, Yersinia enterocolitica, Flavobacterium* sp.) and Gram positive bacteria (*Listeria monocytogenes*) were cultured in nutrient agar and allowed to grow at 37°C in microbial culture laboratory. The fungal cultures of *Aspergillus flavus* and *Pencillium notatum* were used for screening antifungal study. The fungal isolates were allowed to grow on a potato dextrose agar (PDA) until they sporulated.

2.7.2. Screening for Antibacterial Activity

The antibacterial activity was assayed by a modification of agar well diffusion method (Roberto *et al.*, 2001). Different concentrations of the extracts were prepared by reconstituting with DMSO. The test organisms were maintained on agar slants were recovered for testing by inoculating into nutrient broth and incubated at 37°C in a shaker at 180 rpm. The culture of each microorganism was inoculated in plates in nutrient agar and spread evenly using sterile glass spreader. Test extracts were incorporated into the wells made by sterile 5mm size borer in media and different concentration of extract (Methanol, ethanol and water) were added and water alone as a control. Plates were incubated at 37 °C and zone of inhibition was observed after 24 h.

2.7.3. Screening for Antifungal Activity

Antifungal activity of all various extracts was studied against two fungal strains by the agar well diffusion method (Shanmugapriya *et al.*, 2011). The fungal isolates were allowed to grow on a potato dextrose agar (PDA) at 25°C until they sporulated. The fungal spores were harvested after sporulation by pouring a mixture of sterile distilled water. The fungal spore suspension was evenly spread on plate using sterile glass spreader. Wells were then bored into the agar media using sterile 5 mm cork borer and the wells filled with the solution of the extract and water alone as a control. The plates were allowed to stand on a laboratory bench for 1 h to allow for proper diffusion of the extract into the media. Plates were incubated at 25°C for 96 h and later observed for zones of inhibition.

3. Results and Discussions

3.1. Qualitative Analysis of Phytochemicals

Preliminary phytochemical screening of the plant showed the presence of carbohydrate, alkaloids, flavonoids, terpenoids, cardiac glycosides, steroids, tannin, saponin, phlobatinin, glycosides, coumarine, phenol, resin, anthraquinone, amino acids and the absence of acidic compounds and gum/mucilage in fresh samples. The qualitative analysis in fresh samples of *G.polycaulon* plant was tabulated (Table 5.1).

A variety of herbs and herbal extracts contain different phytochemicals with biological activity that can be of valuable therapeutic index. Much of the protective effect of fruits and vegetables has been attributed by phytochemicals, which are the non-nutrient plant compounds. Different phytochemicals have been found to possess a wide range of activities that may help in protection against chronic diseases. The phytochemicals such as saponins, terpenoids, flavonoids, tannins, steroids and alkaloids have anti-inflammatory effects (Kumar *et al.*, 2008). Phenolics are one of the major groups of phytochemical that can be found ubiquitously in certain plants. Flavonoids are capable of treating certain physiological disorder and diseases (Okwu, 2004). Tannins are known to be useful in the treatment of inflamed or ulcerated tissues and they have remarkable activity in cancer prevention and anticancer (Scalbert *et al.*, 1991). Saponin is glycosides occurring widely in plants and medicinally as antibiotic (Oakenfull *et al.*, 1981), antiviral (Okubo *et al.*, 1994), anti-inflammatory (Just *et al.*, 1998) and anti-ulcer. Alkaloids have been associated with medicinal uses for centuries and one of their common biological properties is their cytotoxicity (Harborne JB, 1998).

Table 5.1: Qualitative analysis in fresh samples of *Gnaphalium polycaulon* plant.

Sl.No.	Phytochemicals	Fresh Leaves			Fresh Stem			Fresh Flower		
		M	E	W	M	E	W	M	E	W
1.	Carbohydrates	−	+	−	+	+	−	−	−	−
2.	Alkaloids	+	+	+	+	+	+	+	+	+
3.	Flavonoids	+	+	−	+	−	+	+	+	−
4.	Terpenoids	++	++	++	−	−	++	+	−	−
5.	Cardiac glycosides	−	+	−	−	−	−	−	−	−
6.	Steroids	−	+	−	−	+	−	−	+	−
7.	Tannin	+	−	−	−	−	+	−	+	−
8.	Saponin	+	+	+	+	+	+	+	+	+
9.	Phlobatinin	−	−	−	−	−	+	−	−	+
10	Gum/mucilage	−	−	−	−	−	−	−	−	−
11	Glycosides	+	+	−	−	+	−	+	+	+
12	Coumarine	−	−	−	+	−	+	+	+	+
13	Acidic compounds	−	−	−	−	−	−	−	−	−
14	Phenol	++	++	−	−	−	−	++	+	−
15.	Resin	−	−	−	−	+	−	−	+	−
16.	Anthraquinone	−	−	−	−	+	−	−	−	−
17.	Aminoacids	+	+	+	+	+	+	+	+	+

M: Methanol; E: Ethanol; W: Water.

3.2. Quantitative Analysis of Phytochemicals

The phytochemical screening and quantitative estimation of the percentage crude yields of chemical constituents of the plants studied showed that the leaves and stems were rich in alkaloids, flavonoids, tannins and saponins. They were known to show medicinal activity as well as exhibiting physiological activity (Sofowara, 1993). Flavonoids have been shown to exhibit their actions through effects on membrane permeability, and by inhibition of membrane-bound enzymes such as the ATPase and phospholipase A2 (Lee *et al.*, 2003). The amount of Flavonoids in the fresh stem, leaf and flower of *G. polycaulon* was estimated to be 0.692, 0.713 and 0.698 respectively. The quantitative analysis of *G. polycaulon* plant was reported in Table 5.2.

Tannins are known to be useful in the treatment of inflamed or ulcerated tissues and they have remarkable activity in cancer prevention and anticancer (Scalbert *et al.*, 1991). Thus, Gnaphalium plant containing this compound may serve as a potential source of bioactive compounds in the treatment of cancer. In *G. polycaulon* plant, the maximum amount of tannin was found in the methanolic extract of fresh stem and it was estimated to be 0.48. Saponins are glycosides occurring widely in plants. They are abundant in many foods consumed by animals and man. In medicine, it is used as antibiotic (Oakenfull *et al.*, 1981), antiviral (Okubo *et al.*, 1994), anti-inflammatory

(Just *et al.*, 1998) and anti-ulcer. The amount of Saponin in the fresh stem, leaf and flower of *G.polycaulon* was estimated to be 0.22, 0.12 and 0.18 respectively.

Table 5.2: Quantitative analysis of the *G. polycaulon* plant.

Sl.No.	Quantitative Assay	Fresh Samples (µg/ml)		
		Leaf	Stem	Flower
1.	Total Phenolic content	0.60±0.02	0.16±0.01	0.13±0.01
2.	Total Tannin content	0.42±0.01	0.36±0.02	0.13±0.01
3.	Total Alkaloid content	0.25±0.01	0.19±0.01	0.12±0.02
4.	Total Flavonoid content	0.71±0.03	0.69±0.02	0.49±0.02
5.	Total Saponin content	0.32±0.01	0.22±0.01	0.18±0.01

Quantitative analysis expressed as mean±S.D.

The presence of phenolic compounds contributed to their antioxidative properties and thus the usefulness of plants in herbal medicament. Phenols have been found to be useful in the preparation of some antimicrobial compounds such as dettol and cresol (Olayinka *et al.*, 2010). In *G. polycaulon*, the maximum amount of phenol was found in the methanolic extract of fresh leaf and it was estimated to be 0.60. Alkaloids have been associated with medicinal uses for centuries and one of their common biological properties is their cytotoxicity, and their absence in this plant tend to lower the risk of poisoning by the plant. In *G. polycaulon*, the maximum amount of alkaloid was found in the methanolic extract of fresh flower and it was estimated to be 0.22.

3.3. Thin Layer Chromatography

The appearance of green color spots in leaf extract on the TLC plate indicated the presence of alkaloids, flavonoids, saponins and phenol and *Rf* values were found to be 0.43, 0.49, 0.45 and 0.76. The thick green color spots indicated the presence of glycosides around 0.93. The appearance of green color spots in the stem extract on the TLC plate indicated the presence of alkaloids and glycosides and *Rf* values were found to be 0.39 and 0.69. The appearance of yellow color spots indicated the presence of flavonoids (0.43) and phenol (0.76) whereas the appearance of the pale yellow spots indicated the presence of saponins (0.50) and steroids (0.43). In the flower extract, the appearance of green color spots on the TLC plate indicated the presence of saponins and flavonoids and the *Rf* values were found to be 0.27 and 0.38. The yellow color spots indicated the presence of alkaloids (0.39), steroids (0.32) and phenol (0.73) whereas the pale green spots indicated the presence of flavonoids (0.38).The *Rf* values of different parts of the *G. polycaulon* plant was tabulated in Table 5.3.

3.4. Antimicrobial Activity

The potential for developing antimicrobials from higher plants appears rewarding as it will lead to the development of phytomedicine to act against microbes (Shiwaiker *et al.*, 2006). The antibacterial activity of all solvent extracts of *G. polycaulon* plant was evaluated. The extracts were screened for activity against Gram negative bacteria (*Salmonella typhimurium*, *Yersinia enterocolitica* and *Flavobacterium* sp.) and

Table 5.3: The *Rf* values of different parts of the *G. polycaulon* plant.

Tests	Rf value in cm		
	Leaf	Stem	Flower
Alkaloids	0.43	0.39	0.39
Flavonoids	0.49	0.43	0.38
Saponins	0.51	0.50	0.27
Glycosides	0.93	0.69	0.66
Phenol	0.76	0.76	0.73
Steroids	0.57	0.43	0.32

Table 5.4: Antimicrobial activity of fresh samples of *G. polycaulon* plant.

Microorganisms	Solvents	Zone of Inhibition in mm									Standard
		Fresh Leaf (µg/ml)			Fresh Stem (µg/ml)			Fresh Flower (µg/ml)			
		50	100	150	50	100	150	50	100	150	
Bacteria	Gent										
Salmonella	Methanol	2	2	7	1	3	5	1	2	2	18
typhimurium	Ethanol	2	1	4	1	2	3	–	1	1	
	Water	–	2	6	–	–	4	–	1	1	
Yersinia	Methanol	2	2	4	2	2	2	1	2	2	24
enterocolitica	Ethanol	1	1	2	1	1	2	–	1	1	
	Water	–	–	2	–	1	1	–	1	1	
Flavobacterium sp.	Methanol	2	2	4	1	1	2	1	2	2	20
	Ethanol	2	2	2	1	1	2	1	1	1	
	Water	–	–	1	–	1	1	–	2	2	
Listeria	Methanol	2	2	5	2	4	3	1	2	2	32
monocytogenes	Ethanol	1	2	4	1	3	2	1	1	1	
	Water	–	–	2	–	2	2	–	–	1	
Fungus	Nyst										
Aspergillus flavus	Methanol	2	3	3	2	2	2	1	1	1	08
	Ethanol	2	2	2	1	1	1	–	1	1	
	Water	–	1	2	–	–	1	–	1	1	
Penicillium notatum	Methanol	4	5	6	2	3	4	1	2	2	12
	Ethanol	1	3	4	1	2	2	1	1	2	
	Water	–	–	2	–	1	1	–	–	1	

Gent: Gentamycin; Nyst: Nystatin.

Gram positive bacteria (*Listeria monocytogenes*) by agar well diffusion method. Results were compared with the standard drugs such as gentamycin for bacterial cultures. The zone of inhibition was seen in all extract against all cultures but the maximum inhibition shown by fresh leaf extract is 29 mm respectively. The zone of inhibition of all various fresh extracts of *G.polycaulon* plant was measured and tabulated (Table 5.4). Gentamycin have a universal activity against the entire test organism, with zones of inhibition ranging from 18to 32 mm respectively.

Fungi can cause damage to the structures, decoration of buildings and are also responsible for their indoor air quality (Verma *et al.*, 2011). The antifungal activity of all extracts of *G.polycaulon* plant parts was evaluated using agar well diffusion method. The extracts were tested against *Aspergillus flavus* and *Pencillium notatum*. The extract exhibited significant activity against all the tested fungi compared with the standard drug, Nystatin (10 µg/disc) ranging from 8 to 12 mm respectively. The maximum inhibition shown by fresh leaf extract is 20 mm respectively. All extracts showed good activity against the fungal isolates with zones of inhibition ranging from 8 to 18 mm. In conclusion, the results showed that the all various extract of *G.polycaulon* plant is a broad spectrum agent which can be used against both gram positive and gram negative bacteria and also fungi.

4. Conclusion

Plants contain thousands of constituents and are valuable sources of new and biologically active molecules possessing antimicrobial property. Many reports are available regarding phytochemical, anti-bacterial and anti-fungal, properties of plants. At present, scientists are investigating for plant products of antimicrobial properties. The result showed that *G. polycaulon* plant is good source of phytochemicals. The inhibition produced against particular microorganism depends upon the members of phenolic compounds present and identified by TLC analysis. The antimicrobial study showed maximum inhibition by fresh leaf extract results in good activity against bacterial and the fungal isolates. Many plants with strong therapeutic, medicinal, aromatic and aesthetic effect lie unexplored or remain under explored. There need to be systematic studies to identify and scientifically validate such compounds, which can results in ways of combating the serious disease in the world. These findings help to identify the active components responsible for the development of drugs for therapeutic uses as traditional medicine and suggest that some of the plant extracts possess compounds with good antibacterial properties that can be used as antimicrobial agents in the search of new drugs. Further studies are on going to isolate, identity, characterize and elucidate the structure of the bioactive components.

References

Ashebir, M. and Ashenati, M., 1999. Assessment of the antibacterial activity of some traditional medicinal plants on some food-borne pathogen. *Ethiopian Journal of Health Development*, 13: 211–216.

Azoro, C., 2002. Antibacterial activity of crude extract of *Azudirachita indica* on *Salmonella typhi*. *World Journal of Biotechnology*, 3(1): 347–351.

Benkeblia, N., 2004. Antimicrobial activity of essential oil extract of various onions (*Allium cepa*) and garlic (*Allium sativum*). *Lebensm Wiss-U-Technology*, 37: 263–268.

Bhagwati, U. and Vandana, S., 2005. Traditional knowledge on medicinal plants among rural women of the Garhwal Himalaya, Uttaranchal. *Indian Journal of Traditional Knowledge*, 4(3): 259–266.

Bhupendra, S.A. and Mani, M.B., 2008. Floral diversity of Baanganga Wetland, Uttarakhand, India. *Check List*, 4(3): 279–290.

Calixto, J.B., 2000. Efficacy, safety, quality control, marketing and regulatory guideline for herbal medicines (Phytotherapeutic agents). *Brazil Journals of Medical and Biological Research*, 33: 179–189.

Calixto, J.B., Santos, A.R.S, Filho, V.C. and Tunes, R.A., 1998. A review of the plants of the genus *Phyllanthus*: Their chemistry, pharmacology, and therapeutic potential. *Journal Medical Biological Research*, 31: 225–258.

Colombo, M.L. and Bosisio, E., 1996. Pharmacological activities of *Chelidonium majus* L. Paparveraeae). *Pharmacological Resources*, 33: 127–134.

Cowan, M.M., 1999. Plant Products as antimicrobial agents. *Clinical Microbiology Reviews*, 2(4): 564–582.

Harborne, J.B., 1973. *Phytochemical Methods*. Chapman and Hall, Ltd., London, pp. 49–188.

Hufford, C.D., Jia, Y., Croom, E.M., Muhammed, I., Okunade, A.L., Clark, M. and Rogers, R.D., 1993. Antimicrobial compounds from *Petalostemum purpureum*. *Journal of Natural Products*, 56: 1878–1889.

Just, M.J., 1998. Anti-inflammatory activity of unusual lupane saponins from *Buleurum fruiticescens*. *Planta Medica*, 64(5): 404–407.

Kumar, S., Kumar, D., Manjusha, S.K., Singh, N. and Vashishta, B., 2008. Antioxidant and free radical scavenging potential of *Citrullus coloctnthis*(L) schrad methanolic fruit extract. *Acta Pharmacology*, 58: 215–220.

Le, S.R., 1977. *A History of Herbal Plants*, 4th edn. Angus and Robertson Limited. London, pp. 50–55.

Lee, S.E., Hwang, H.J. and Ha, J.S., 2003. Screening of medicinal plant extracts or antioxidant activity. *Life Sciences*, 73: 167–179.

Oakenfull, Fenwick, 1981. Saponin content of soybeans and some commercial soybean products. *Journal of Food Agricultural Science*, 32: 273–278.

Obadoni, B.O. and Ochuko, P.O., 2001. Phytochemical studies and comparative efficacy of crude extracts of some homeostatic plants in Edo and Delta States of Nigeria. *Global Journal of Pure Applied Sciences*, 8: 203–208.

Obdoni, B.O. and Ochuko, P.O., 2001. Phytochemical studies and comparative efficacy of the crude extracts of some Homostatic plants in Edo and Delta States of Nigeria. *Global Journal Pure Applied Science*, 8: 203–208.

Odebiyi, E.O. and Ramstard, A.H., 1978. Investigation photochemical screening and antimicrobial screening of extracts of *Tetracarpidium conophorum*. *Journal of Agricultural Chemical Society*, 26: 1–7. ACS Symposium Series.

Ok wu, D.E., 2004. Phytochemical and Vitamin content of indigenous species of South Eastern Nigeria. *Journal of Sustainable Agriculture and the Environment*, 6: 30–34.

Olayinka, A.A. and Okoh, A., 2010. Preliminary phytochemical screening and *In vitro* antioxidant activities of the aqueous extract of *Helichrysum longifolium*. *BMC Complementary and Alternative Medicine*, 10(21): 1–8.

Robbers, J., Speeddie, M. and Tyler, V.,1996. *Pharmacognosy and Pharmacobiotechnology*. Williams and Wilking.Baltimore, pp. 1–14.

Roberto, V.I., Maricruz, S., Ofelia, E., Armida, Z.E., Martýn, T.V. and Pedro, J.N., 2001. Antimicrobial activity of three Mexican *Gnaphalium* species. *Fitoterapia*, 72: 692–694.

Rojas, A. H., Pereda, M.R. and Meta, R., 2003. Sreening of antimicrobial activity of Crude drug extracts and pure natural products from Mexican medicinal plants. *Journal of Ethnopharmacology*, 35: 275–283.

Sandeep, A., 2011. Presage biology: lesson from nature in weather forecasting. *Indian Journal of Traditional Knowledge*, 10(1): 114–124.

Santoyo, S., Jaime, L., Herrero, M., Senorans, F., Cifuentes, A. and Ibanez, E., 2006. Functional characterization of pressurized liquid extracts of *Spirulina platensis*. *European Food Research Technology*, 224: 75–81.

Scalbert, A., 1991. Antimicrobial properties of tannins. *Phytochemistry*, 30: 3875–3883.

Shanmugapriya, K., Saravana, P.S., Harsha, P., Peer, M. and Binnie, W., 2011. A comparative study of antimicrobial potential and phytochemical analysis of *Artocarpus heterophyllus* and *Manilkara zapota* seed extracts. *Journal of Pharmacy Research*, 4(8): 2587–2589.

Shiwaiker, A., Rajendran, K. and Kumar, C.D., 2006. *In vitro* antioxidant studies of Annonasquamosa Linn leaves. *Indian Journal Experimemtal Biology*, 42: 803–807.

Siddhuraju, P. and Becker, K., 2003. Antioxidant properties of various solvent extracts of total phenolic constituents from three different agro climatic origins of Drumstick tree (*Moringa oleifera* Lam.) leaves. *Journal of Agricultural Food Chemistry* 51: 2144–2155.

Sofowora, A., 1993. Screening plants for Bioactive Agents. In: *Medicinal Plants and Traditional Medicines in Africa*, 2nd edn. Spectrum Books Ltd. Sunshine House, Ibadan, Nigeria, pp. 134–256.

Van-Burden, T.P. and Robinson, W.C., 1981. Formation of complexes between protein and Tannin acid. *Journal of Agricultural Food Chemistry*, 1: 77.

Verma, R.K., Chaurasia, L. and Kumar, M., 2011. Antifungal activity of essential oils against selected building fungi. *Indian Journal of Natural Products Resources*, 2(4): 448–451.

Witham, F.H., Balydes, D.F. and Devlin, R.M., 1971. *Experiment in Plant Physiology*. Van Nostrant Company, New York, p. 245.

Xia, L., Yuangang, L., Yujie, F., Liping, Y. and Chengbo, G., 2005. Antimicrobial activity and cytotoxicity towards cancer cells of *Melaleuca alternifolia* (tea tree) oil. *European Food Research and Technology*, 229: 247–253.

Yen, G.C. and Chen, H.Y., 1995. Antioxidant activity of various tea extracts in relation to their antimutagenicity. *Journal of Agricultural Food Chemistry*, 47: 23–32.

Zhishen, J., Mengcheng, T. and Jianming, W., 1999. The determination of flavonoids contents in mulberry and their scavenging effects on superoxide radicals. *Food Chemistry*, 64: 555–559.

Scientific Basis of Herbal Medicine (2013)
Editor: Dr. Parimelazhagan Thangaraj
Published by: DAYA PUBLISHING HOUSE, NEW DELHI

Pages 61–66

Chapter 6

Antibacterial and Antioxidant Properties of Ginger Rhizome Extract

Shiji Wilson and C.K. Padmaja

Department of Botany,
Avinashilingam University, Coimbatore – 43,
Tamil Nadu, India

1. Introduction

In ancient times herbs and spices were used to preserve foods. Their effectiveness in food preservation was the result of their potent antioxidant and antimicrobial properties. Antioxidants are substances that when present in low concentrations, compared to those of an oxidisable substrate significantly delay or prevent oxidation of that substance (Halliwell and Gutteridge, 1989). Plants are a potential source of natural antioxidants or phytochemical. Ginger (*Zingiber officinale* L. Rosc) belongs to the family Zingiberaceae of Monocotyledonae. It is one of the traditional folk medicinal plant that have been used for over 2000 years by Polynesians for treating diabetes, high blood pressure, cancer, fitness and many other illnesses (Tepe *et al.*, 2006). The extract obtained from the root contains polyphenol compounds (6-gingerol and its derivatives) which have a high antioxidant activity (Herrmann, 1994). Ginger (*Zingiber officinale*), is not only one of the most popular of all the spices but is also of the top five antioxidant foods. Several phytochemicals found in ginger have demonstrated strong anticancer activities in both laboratory and clinical studies. Cancer is often associated with inflammatory processes anti-inflammatory activity of ginger reduces the risk of inflammation-induced malignancy. The pungent vanilloids, gingerol and paradol found in ginger, are very effective in killing cancer cells. They achieve this, both by direct cytotoxic activity against the tumor and indirectly by inducing apoptosis in the cancer cells and reducing tumor initiation and growth. Ginger can prevent DNA damage. Melatonin is an antioxidant produced by the body that is also found in some plants, such as ginger. It has the valuable property of being able to access in most parts of the body, including brain and nervous tissue, and protects DNA against carcinogenic free-radical damage.

2. Materials and Methods

2.1. Collection of Materials

Ginger rhizome was collected from the field, shade dried, powdered and was used for the study. Petroleum ether, chloroform, methanol and water extracts was prepared in Soxhlet apparatus.

2.2. Pharmacognostic Study

The extracts were subjected to pharmacognostic studies, which included

1. Antimicrobial activity
2. Antioxidant activity

2.3. Screening of Extracts for Antibacterial Analysis

2.3.1. Test Bacterial Strains

Cultures of *Klebsiella pneumonia*, *Salmonellla* sp. and *Shigella sonnei* were obtained from Pathology Department, PSG Medical College, and Coimbatore and maintained as subculture in slants in the Department of Botany, Avinashilingam Deemed University for Women, Coimbatore.

2.3.2. Culture Media for Bacteria

Nutrient Agar

1000 ml Nutrient agar was prepared by adding Peptone 5.0 g, Sodium chloride 5.0 g, Beef extract 1.5 g, Yeast extract 1.5 g and Agar 15.0 g in 1000 ml distilled water and the pH was adjusted to 7.0.

2.3.3. Antibacterial Studies

Antibacterial studies were carried out by Agar well diffusion and Disc diffusion method. (Bauer *et al.*, 1966)

2.4. Screening for Antioxidant Compounds

2.4.1. Enzymatic Antioxidants Activity

1. Estimation of catalase (Vir and Grewal, 1975)
2. Estimation of peroxidase (Malik and Singh, 1980)
3. Estimation of superoxide dismutase (Das *et al.*, 2000)
4. Estimation of glutathione reductase (Beutler, 1984)
5. Estimation of glutathione peroxidase (Ellman, 1959)

2.4.2. Non-enzymatic Antioxidants Activity

1. Estimation of β-carotene (vitamin A) (Bayfield and Cole, 1974)

2. Estimation of tocopherol (vitamin E) (Rosenberg, 1992)
3. Estimation of ascorbic acid (vitamin C) (Roe and Kuether, 1953)

2.5. Statistical Analysis

The data observed from various antioxidants observations were subjected to statistical analysis and based on the results, inferences were drawn (Panse and Sukhatme, 1978).

3. Results and Discussion

3.1. Antibacterial Activity

The results of antibacterial activity revealed that the maximum inhibition zone of 7.3 mm and 7.1 mm were obtained in water extract in both agar well and disc diffusion method by petroleum extract in agar well diffusion method (6.0 mm) followed by petroleum ether extracts in agar well diffusion method (6.0 mm) against *K. pneumoniae* and in disc diffusion method (5.6 mm) against *Salmonalla* species. The least inhibition zone of 1.0 mm was recorded in petroleum extract in agar well diffusion method and 0.5 mm in chloroform extract in disc diffusion method against *S.sonnei*.

Table 6.1: Inhibition zone by ginger rhizome extract againsttest organisms (agar well diffusion method).

Test Organism	Plant	Zone of Inhibition (mm)				
		Petroleum Ether	Chloroform	Methanol	Water	Control*
Klebsiella pneumoniae	Ginger	6.0	5.6	3.2	7.3	4.2
Salmonella sp.	Ginger	3.0	2.6	2.1	1.2	8.4
Shigella sonnei	Ginger	1.0	5.3	4.6	5.2	6.1

Control*: Chloramphenicol.

Table 6.2: Inhibition zone by ginger rhizome extract against testorganisms (disc diffusion method).

Test Organism	Plant	Zone of Inhibition (mm)				
		Petroleum Ether	Chloroform	Methanol	Water	Control*
Klebsiella pneumoniae	Ginger	2.2	3.1	1.1	7.1	4.0
Salmonella sp	Ginger	5.6	5.4	3.6	2.4	8.2
Shigella sonnei	Ginger	1.4	0.5	2.1	3.2	6.3

Control*: Chloramphenicol.

Similar result was reported by Yusha's *et al.* (2008). They reported that the ethanol extract of ginger showed remarkable activity against all the organisms (*K. pneumoniae, Pseudomonas aeruginosa, Escherichia coli* and *Proteus vulgaris*) with highest inhibition zone of 35 mm on *E. coli*.

The present result is in accordance with the result of Gao and Zhang (2010). They also found that the alcohol extract of ginger rhizome exhibited inhibition zone of 9.27 mm against *Shigella castellani.*

3.2. Antioxidants

Antioxidants are molecules which can safely interact with free radicals and terminate the chain reaction before vital molecules are damaged. The antioxidants are classified as enzymatic and non- enzymatic antioxidants. The important enzymatic antioxidants are catalase, peroxidase, superoxide dismutase, glutathione peroxidase and glutathione reductase, and non enzymatic antioxidants are ascorbic acid (Vitamin C), tocopherol (Vitamin E) and β-carotene (Vitamin A).

3.2.1. Enzymatic Antioxidants (Table 6.3)

Among the enzymatic antioxidants, superoxide dismutase activity was much pronounced in ginger rhizome (19.74 units) followed by peroxidase (8.44 units) and glutathione peroxidase (8.09 units). The least activity of (0.59 units) was expressed in catalase activity.

The present result is in accordance with the findings of Sheela and Ramani (2011). *In vitro* antioxidant activity of *Polygrown barbatum* leaf extract was investigated by different methods. DPPH, radical scavenging assay, nitricoxide scavenging assay, superoxide scavenging assay and total phenolic content. Thus the *in vitro* studies indicate the ethanolic extract has significant antioxidant activity and also a better source of natural antioxidant.

Table 6.3: Enzymatic antioxidants.

Enzymatic Antioxidants in units (u)	Ginger	SED	C.D P<0.01
Catalase*	0.59	0.01	0.06
Peroxidase**	8.44	0.50	2.29
Superoxide dismutase***	19.74	0.74	3.42
Glutathione reducates****	6.94	0.61	2.80
Glutathione proxidase*****	8.09	0.10	0.45

*   Catalase activity (Units) = Equivalent to 1 ml of 0.01 N potassium permanganate utilized by the enzyme min^{-1} mg^{-1} enzyme protein

**   Peroxidase activity (Units) = μ mols of pyrogalol oxidized min^{-1} mg^{-1} enzyme protein

*** Superoxide dismutase (Units) = μ mols of NADPH reduced min^{-1} mg^{-1} enzyme protein

**** Glutathione reductase (Units) = μ mols of NADPH oxidized min^{-1} mg^{-1} enzyme protein

***** Glutathione Peroxidase (Units) =μg of glutathione consumed min^{-1} mg^{-1} enzyme protein.

Statistical investigation of the results also revealed that among the enzymatic antioxidants, superoxide dismutase, was significantly much pronounced in ginger rhizome (P<0.01) than the other enzymes.

3.2.2. Non-Enzymatic Antioxidants (Table 6.4)

The non-enzymatic antioxidants are ascorbic acid (Vitamin C), tocopherol (Vitamin E) and β-carotene (Vitamin A). Vitamins are small molecule antioxidant derived exclusively from the diet. They have tremendous role from medicinal point of view.

Tocopherol (Vitamin E) was found to be significantly at a higher level of 7.39 µg g^{-1} followed by β–Carotene (Vitamin A) showed 4.18 µg g^{-1} in ginger rhizome and ascorbic acid (Vitamin C) was found to be at least amount (0.34 µg g^{-1}).

Table 6.4: Non-enzymatic antioxidants.

Non-Enzymatic Antioxidants	Ginger (µg g^{-1})	SED	C.D
β-caroteine (Vitamin A)	4.18	0.19	0.87
Ascorbic acid (Vitamin C)	0.34	0.02	0.08
Tocopherol (Vitamin E)	7.39	0.45	2.06

The present result is in conformity with the findings of Kamala (2006) who also obtained tocopherol (Vitamin E) and ascorbic acid (Vitamin C) in the range from 5.00 to 10.00 µg g^{-1} and from 0.19 to 0.29 µg g^{-1}.

4. Conclusion

Thus, it can be inferred from present findings that the ginger rhizome posses both antibacterial and antioxidant property. The antibacterial and antioxidant properties acts as promising pharmacological agent for preventing cancer, cardiovascular diseases, inflammatory disorders, neurological degenerations, wound healing, infectious disease and aging. Ginger is also be used as a food preservative.

References

Bauer, A.W., Kirley, M.D.K., Sherris, T.C. and Track, M., 1966. Antibiotic susceptibility testing by standardized single disc diffusion method. *American J. Cli. Pathol.*, 45: 493–496.

Bayfield, H.T. and Cole, E.A., 1974. Calorimetric determination of Vitamin A with trichloroacetic acids. *Methods Enzymol.*, 67: 189–195.

Beutler, E., 1984. *Red Cell Metabolism, Manual of Biochemical Methods*, 3rd edn. Grune Stra © on, Inc. Orlando, FL 32887, London.

Das, S. and Vasisht, S., 2000. Correlation between total antioxidant status and lipid peroxidation in hypercholersterolemia. *Curr. Sci.*, 78: 486.

Ellman, G., 1959. Tissue sulpfhydryl groups. *Archiv. Biochem. and Biophy.*, 32: 70–77.

Gao, D. and Zhang, Y., 2010. Comparative antibacterial activities of crude polysaccharides and flavonoids from *Zingiber officinale* and their extraction. *Nat. Prod. Res. Dev.*, 21(10): 459–461.

Halliwell, B. and Gutteridge, J.M.C., 1989. *Free Radicals in Biology and Medicine*, 2nd edn. Clarendon Press, Oxford, UK.

Herrmann, K., 1994. Antioxidativ wiksame Pflanzenphenole sowie Carotinoide als wichtige Inhaltsstoffe von Gewu¨rzen. *Gordian*, 94: 113–117.

Kamala-Raj, 2006. Betel leaf: The neglected green gold of India (P. Guha) Agricultural and Food Engineering Department, Indian Institute of Technology, Kharagpur, West Bengal, India. *J. Hum. Ecol.*, 19(2): 87–93.

Malik, C.P. and Singh, M.B., 1980. *Plant Enzymology and Hitoenzymology.* Kalyani Publishers, New Delhi, p. 53.

Panse, V.G. and Sukhatme, P.V., 1978. *Statistical Methods for Agricultural Workers.* ICAR Publ., New Delhi, p. 361.

Rosenberg, H.R., 1992. *Chemistry and Physiology of Vitamins.* International Science Publishers Inc., NewYork, pp. 452–453.

Roe, J.H. and Kuether, C.A., 1953. The determination of ascorbic acid in whole blood and urine through 2,4-dinitrophenyl and hytrazine derivative dehydro ascorbic acid. *J. Biol. Chem.*, 147: 399–407.

Sheela Q.R. and Ramani A.V., 2011. *In vitro* antioxidant activity of *Polygonum barabatum* leaf extract. *Asian J. Pharm. Clin. Res.*, 4(1): 113–115.

Tepe, B., Sokmen, M., Akpulat, H.A. and Sokmen, A., 2006. Screening of the antioxidant potentials of six *Salvia* species from Turkey. *Food Chem.*, 95: 200–204.

Vir, S. and Grewal, J.S., 1975. Change in the catalase activity of gram plant induced by *Ascochla rabiei* infection. *Indian Phyto. Pathol.*, 28: 223–225.

Yusha's, M., Garba, L. and Shamsuddeen, U., 2008. *In vitro* inhibitory activity of garlic and ginger extracts on some respiratory tract isolates of gram-negative organisms. *Int. J. Biomed. Hlth. Sci.*, 4(2): 57–60.

Scientific Basis of Herbal Medicine (2013)
Editor: Dr. Parimelazhagan Thangaraj
Published by: DAYA PUBLISHING HOUSE, NEW DELHI

Pages 67–72

Chapter 7

Mosquito Larvicidal Activity of Jasmine Oil (*Jasminum sambac* L.) against *Culex tritaeniorhynchus* (Diptera: Culicidae)

K. Yogalakshmi, M. Govindarajan,
R. Sivakumar and M. Rajeswary

*Division of Vector Biology and Phytochemistry,
Department of Zoology, Annamalai University,
Annamalai Nagar – 608 002, Tamil Nadu, India*

1. Introduction

Mosquito-borne diseases have an economic impact, including loss in commercial and labor outputs, particularly in countries with tropical and subtropical climates; however, no part of the world is free from vector-borne diseases. Mosquitoes are the major vector for the transmission of malaria, dengue fever, yellow fever, filariasis, schistosomiasis and Japanese encephalitis (JE). Mosquitoes also cause allergic responses in humans that include local skin and systemic reactions such as angioedema (Peng *et al.*, 1999). *Culex tritaeniorhynchus* Giles is an important vector of JE in India and South East Asian countries. JE is endemic in few states of India and highly endemic in few districts of Tamil Nadu, Southern India (Reuben and Gajanana, 1997). Keiser *et al.* (2005) have reported that approximately 1.9 billion people currently live in rural JE prone areas of the world, the majority of them in China (766 million) and India (646 million). Synthetic insecticides have created a number of ecological problems, such as the development of resistant insect strains, ecological imbalance and harm to mammals. Hence there is a constant need for developing biologically active plant materials as larvicides, which are expected to reduce the hazards to human and other organisms by minimizing the accumulation of harmful residues in the environment. The essential oils and extracts of edible and medicinal plants, herbs, and spices constitute a class of very potent natural bioactive compounds used by cosmetics, pharmaceutical, and food industries. Thus, many researchers were intrigued to exploit essential oils as a potential source for the identification of novel natural pest control agents, with a strong focal point on mosquito control (Pushpanathan *et al.*, 2008).

Govindarajan (2011) have been reported that the larvicidal and repellent properties of essential oil from various parts of four plant species namely *Cymbopogan citrates, Cinnamomum zeylanicum, Rosmarinus officinalis* and *Zingiber officinale* against *Culex tritaeniorhynchus* and *Anopheles subpictus*. Essential oils provide a rich source of biologically active monoterpenes and are well documented for bioactivities against insect pests. Some of the essential oils with promising mosquito control potential are from the genus *Tagetes* spp. (Vasudevan *et al.*, 1997), *Ocimum* spp. (Bhatnagar *et al.*, 1993) and *Cymbopogon* spp. (Ansari and Razdan 1994) etc. Further, essential oils of *cassia, camphor, wintergreen, pine*, and *eucalyptus* are already being used in several commercial products for mosquito control (Ansari and Razdan, 1995). Essential oils of leaf and bark of *Cryptomeria japonica* demonstrated high larvicidal activity against *A. aegypti* larvae (Cheng *et al.*, 2003). The essential oil of *Z. officinalis* as a mosquito larvicidal and repellent agent against the filarial vector *C. quinquefasciatus* (Forty essential oils extracted from Australian plants were evaluated against mosquitoes, march flies, and sand flies. The most effective of these were *Dacrydium franklini, Backhousia myrtifolia, Melaleuca bracteata*, and *Zieria smithii* (Penfold and Morrison 1952). The present study was an attempt to assess larvicidal properties of the Jasmine oil (*Jasminum sambac*) against early third larvae of *Cx. tritaeniorhynchus*.

2. Materials and Methods

2.1. Plant Material and Extraction of Essential Oil

Jasminum sambac flower was commercially purchased and it was authenticated by a plant taxonomist from the Department of Botany, Annamalai University. A voucher specimen is deposited at the herbarium of plant phytochemistry division, Department of Zoology, Annamalai University. Essential oil was obtained by the hydro- distillation of 2.5 kg fresh flowers in a Clevenger apparatus for 8 h. The oil layer was separated from the aqueous phase using a separating funnel. The resulting essential oil was dried over anhydrous sodium sulphate and stored in an amber-coloured bottle at 8°C for analysis.

2.2. Test Organisms

The mosquito *Culex tritaeniorhynchus* was reared in the vector control laboratory, Department of zoology, Annamalai University. The larvae were fed on dog biscuits and yeast powder in the 3: 1 ratio. Adults were provided with 10 per cent sucrose solution and one week old chick for blood meal. Mosquitoes were held at 28±2°C temperature, 70–85 per cent relative humidity (RH), with a photo period of 14 h light, 10 h dark.

2.3. Larvicidal Activity

Larvicidal activity of the Jasmine oil (*Jasminum sambac*) was evaluated according to WHO protocol (2005). Based on the wide range and narrow range tests, essential oil was tested at 100, 200, 300, 400 and 500 ppm. Essential oil was dissolved in 1 ml

DMSO, and then diluted in 249 ml of filtered tap water to obtain each of the desired concentrations. The control was prepared using 1 ml of DMSO in 249 ml of water. Twenty late third instar larvae were then introduced into each solution. For each concentration, five replicates were performed, for a total of 100 larvae. Larval mortality was recorded at 24 h after exposure, during which no food was given to the larvae. The lethal concentrations (LC_{50} and LC_{90}) were calculated by probit analysis (Finney, 1971).

2.4. Statistical Analysis

The average larval mortality data were subjected to probit analysis for calculating LC_{50}, LC_{90} and other statistics at 95 per cent confidence limits of upper confidence limit and lower confidence limit, and Chi-square values were calculated using the SPSS 12.0 (Statistical Package of Social Sciences) software. Results with $p<0.05$ were considered to be statistically significant.

3. Results and Discussion

The efficacy of Jasmine oil (*Jasminum sambac*) was tested against the early third larvae of *C. tritaeniorhynchus*. The data were recorded and statistical data regarding the LC_{50}, LC_{90}, Chi-square and 95 per cent confidence limits were calculated (Table 7.1). The LC_{50} and LC_{90} values of jasmine oil against early third larvae of *C. tritaeniorhynchus*, were 451.70 and 800.79 ppm, respectively. No mortality was observed in control. The chi-square value were significant at p < 0.05 level.

Table 7.1: Larvicidal activity of Jasmine oil (*Jasminum sambac*) against *Culex tritaeniorhynchus*.

Name of the Oil	Concentration (ppm)	Per cent of mortality±SD	LC_{50} (ppm) (LCL-UCL)	LC_{90} (ppm) (LCL-UCL)	χ^2
Jasmine oil	control	0.0±0.0	245.7	476.79	12.420*
(*Jasminum sambac*)	100	18.9±0.5	(199.49- 291.04)	(410.42- 558.65)	
	200	43.1±0.7			
	300	61.3±0.4			
	400	82.1±1.0			
	500	99.5±0.6			

* Significant at P<0.05.

SD: Standard Deviation; LCL: Lower Confidence Limits; UCL: Upper Confidence Limits; χ^2: Chi square.

Plant essential oils, in general, have been recognised as an important natural resource of insecticides. The results of the present study are also comparable to the earlier reports on the larvicidal activities of plant essential oils. The essential oil of *Ipomoea cairica* showed 100 per cent mortality at 170 ppm for *C. tritaeniorhynchus* and 120 and 170 ppm for *A. aegypti* and *A. stephensi*, respectively (Thomas *et al.*, 2004). The larvicidal activity of essential oils of *Citrus sinensis*, *Eucalyptus* spp., *Ferrula hermonis*,

Laurus nobilis and *Pinus pinea* against *C. pipiens* has LC_{50} values of 60.0, 120.0, 44.0, 117.0 and 75.0 ppm, respectively (Traboulsi *et al.*, 2005). Tiwary *et al.* (2007) observed the larvicidal activity of linalool rich essential of *Zanthoxylum armatum* against different mosquito species *viz.*, *C. quinquefasciatus* (LC_{50}=49 ppm), *A. aegypti* (LC_{50}=54 ppm) and *A. stephensi* (LC_{50}=58 ppm). Cheng *et al.* (2003) examined plant essential oils against *A. aegypti* larvae with LC_{50} values ranging from 36.0 to 86.8 µg/ml. Cavalcanti *et al.* (2004) reported that the larvicidal activity of essential oils from Brazilian plants with LC_{50} values ranging from 60 to 69 µg/ml against *A. aegypti* larvae. Rahuman *et al.* (2000) also found that n-hexadecanoic acid in *Feronia limonia* dried leaves was effective against fourth-instar larvae of *C. quinquefasciatus*, *A. stephensi* and *A. aegypti* with LC_{50} values of 129.24, 79.58, and 57.23 µg/ml, respectively. The essential oil from the leaves of *C. anisata* exhibited significant larvicidal activity, with 24 h LC_{50} values of 140.96, 130.19 and 119.59 ppm, respectively (Govindarajan 2010).

The highest larvicidal activity was observed in the essential oil from *Z. officinale* against *C. tritaeniorhynchus* and *An. subpictus* with the LC_{50} and LC_{90} values as 98.83, 57.98 ppm and 186.55, 104.23 ppm, respectively (Govindarajan 2011). Singh *et al.* (2006) presents the mosquito larvicidal property of *M. charantia* fruit against *An. stephensi* (LC_{50} value 66.05 ppm) and *C. quinquefasciatus* (LC_{50} value 96.11 ppm). Batabyal *et al.* (2007) reported larvicidal responses of seed extracts of plants, *Azadirachta indica* (LC_{50} of 131.32 ppm), *Ricinus communis* (LC_{50} of 194.98 ppm), and *M. charantia* (LC_{50} of 87.00 ppm) to *A. stephensi* larvae. Mohan *et al.* (2005) analyzed the larvicidal effect of *Solanum xanthocarpum* fruit extracts against *A. stephensi* (LC_{50} of 28.62 and 26.09 ppm) and *C. quinquefasciatus* (LC_{50} of 62.62 and 59.45 ppm) after 24 and 48 h of exposure respectively. Essential oils from *Z. officinalis* was evaluated for larvicidal and repellent activity against the filarial mosquito *C. quinquefasciatus*. The LC_{50} value was 50.78 ppm, Skin repellent test at 1.0, 2.0, 3.0, and 4.0 mg/cm^2 concentration of *Z. officinalis* gave 100 per cent protection up to 15, 30, 60, and 120 min (Pushpanathan *et al.*, 2008). Compared with earlier reports, our results revealed that the experimental essential oil was effective to control *C. tritaeniorhynchus*. From these results it was concluded that the plant Jasmine oil (*J. sambac*) exhibits larvicidal activity against important vector mosquito.

4. Conclusion

The widespread use of synthetic organic insecticide during the last five decades has resulted in environment hazards and development of resistances in the major vector species. This has necessitated the search and development of environmentally safe, biodegradable, low-cost, and indigenous methods for vector control, which can used with minimum care, and indigenous method for vector control, which can used with minimum care by individuals and communities in specific situations. In our result, it showed that jasmine oil have significant larvicidal activity. Further studies on identification of active compounds, toxicity and field trials are needed to recommend the essential of this plant for development of eco-friendly chemicals for control of insect vectors.

References

Ansari, MA. and Razdan, R.K., 1994. Repellent action of *Cymbopogon martini martini* Stapf var. *Sofia* against mosquitoes. *Indian J. Malariol.*, 31: 95–102.

Ansari, M.A. and Razdan, R.K., 1995. Relative efficacy of various oils in repelling mosquitoes. *Indian J. Malariol.*, 32: 104–111.

Batabyal, L., Sharma, P., Mohan, L., Maurya, P. and Srivastava, C.N., 2007. Larvicidal efficiency of certain seed extracts against *Anopheles Stephensi*, with reference to *Azadirachta indica*. *J. Asia Pacific Entomol.*, 10: 1–5.

Bhatnagar, M., Kapur, K.K., Jalers, S. and Sharma, S.K., 1993. Laboratory evaluation of insecticidal properties of *Ocimum basilicum* L. and *Ocimum sanctum* L. plants essential oils and their major constituents against vector mosquito species. *J. Entomol. Res.*,17: 21–26.

Cavalcanti, E.S.B., Morais, S.M., Lima, M.A.A. and Santana, E.W.P., 2004. Larvicidal activity of essential oils from Brazilian plants against *Aedes aegypti* L. *Mem. Inst. Oswaldo Cruz*, 99: 541–544.

Cheng, S.S., Chang, H.T., Chang, S.T., Tsai, K.H. and Chen, W.J., 2003. Bioactivity of selected plant essential oils against the yellow fever mosquito *Aedes aegypti* larvae. *Bioresour. Technol.*, 89: 99–102.

Finney, D.J., 1971. *Probit Analysis*. Cambridge University Press.

Govindarajan, M., 2010. Chemical composition and larvicidal activity of leaf essential oil from *Clausena anisata* (willd.) Hook. F. Benth (Rutaceae) against three mosquito species. *Asian Pacific J. Trop. Med.*, 3: 874–877.

Govindarajan, M., 2011. Larvicidal and repellent properties of some essential oils against *Culex tritaeniorhynchus* Giles and *Anopheles subpictus* Grassi (Diptera: Culicidae). *Asian Pacific J. Trop. Med.*, 4(2): 106–111.

Keiser, J., Maltese, M.F., Erlanger, T.E., Bos, R., Tanner, M., Singer, B.H. and Utzinger, J., 2005. Effect of irrigated rice agriculture on *Japanese encephalitis*, including challenges and opportunities for integrated vector management. *Acta Tropica*, 95: 40–57.

Mohan, L., Sharma, P. and Srivastava, C.N., 2005. Evaluation of *Solanum xanthocarpum* extracts as mosquito larvicides. *J. Environ. Biol.*, 26: 399–401.

Penfold, A.R. and Morrison, F.R., 1952. Some Australian essential oils in insecticides and repellents. *Soap. Perfum. Cosmet.*, 52: 933–934.

Peng, Z., Yang, J., Wang, H. and Simons, F.E.R., 1999. Production and characterization of monoclonal antibodies to two new mosquito *Aedes aegypti* salivary proteins. *Insect Biochem. Mol. Biol.*, 29: 909–914.

Pushpanathan, T., Jebanesan, A. and Govindarajan, M., 2008. The essential oil of *Zingiber officinalis* Linn (Zingiberaceae) as a mosquito larvicidal and repellent agent against the filarial vector *Culex quinquefasciatus* Say (Diptera: Culicidae). *Parasitol. Res.*, 102: 1289–1291.

Rahuman, A.A., Gopalarkrishnan, G., Saleem, G., Arumrgam, S. and Himalayan, B., 2000. Effect of *Feronia limonia* on mosquito larvae. *Fitoterapia*, 71: 553–555.

Reuben, R. and Gajanana, A., 1997. Japanese encephalitis in India. *Indian J. Pediatr.*, 64: 243–251.

Singh, R.K., Dhiman, R.C. and Mittal, P.K., 2006. Mosquito larvicidal properties of *Momordica charantia* Linn (Family: Cucurbitaceae). *J. Vect. Borne Dis.*, 43: 88–91.

Thomas, T.G., Rao, S. and Lal, S., 2004. Mosquito larvicidal properties of essential oil of an indigenous plant, *Ipomoea cairica* Linn. *Jpn. J. Infect. Dis.*, 57: 176–177.

Tiwary, M., Naik, S.N., Tewary, D.K., Mittal, P.K. and Yadav, S., 2007. Chemical composition and larvicidal activities of the essential oil of *Zanthoxylum armatum* DC (Rutaceae) against three mosquito vectors. *J. Vect. Born Dis.*, 44: 198–204.

Traboulsi, A.F., El-Haj, S., Tueni, M., Taoubi, K., Nader, N.B. and Mrad, A., 2005. Repellency and toxicity of aromatic plant extracts against the mosquito *Culex pipiens molestus* (Diptera: Culicidae). *Pest Manag Sci.*, 61: 597–604.

Vasudevan, P., Kashyap, S. and Sharma, S., 1997. *Tagetes*: a multipurpose plant. *Bioresour. Technol.*, 62: 29–35.

World Health Organization, 2005. *Guidelines for Laboratory and Field Testing of Mosquito Larvicides*. WHO/CDS/WHOPES/GCDPP/2005.13. Geneva, p. 69.

Scientific Basis of Herbal Medicine (2013)
Editor: Dr. Parimelazhagan Thangaraj
Published by: DAYA PUBLISHING HOUSE, NEW DELHI

Pages 73–85

Chapter 8

In vitro Micropropagation and Callus Induction of *Cassia tora* L. (Tagarai): A Potential Medicinal Herb

Anju Singh and P. Kumudha

Department of Botany, Avinashilingam University,
Coimbatore – 641 043, Tamil Nadu, India

1. Introduction

Plants have been used medicinally throughout the history of life. Before the beginning of the 19th century, many herbs were considered conventional medicines and were included in medicinal curricula and formularies. It is estimated that nearly 80 per cent of the world's population use plants as drugs. India possesses a rich and diverse variety of plant resources to meet the growing demand for plant based drugs, perfumery and flavour items because of the wide variation in soil and climate. It is estimated that at present there are more than 10, 800 licensed pharmacies in Indian systems of medicine and nearly 4, 60,000 registered practitioners of Ayurveda, Siddha, Unani and Homeopathy medicine. Naturally occurring drugs have been obtained from plants, animals and micro-organisms but the natural products with the broadest range of therapeutic application are obtained from the plant kingdom (Prajapat and Kumar, 2003).

Cassia tora L. of the family Caesalpiniaceae occurs throughout India. It is widely cultivated in Korea, China, Japan, Philippines, Vietnam, Indonesia and North America and it mainly occurs in waste lands as a rainy season weed. The most popular English names of *Cassia tora* are Foetid *Cassia,* Wild senna and Sickle senna. *Cassia tora* L. is an annual, foetid herb with 30-90 cm in height. It grows in dry soil throughout tropical parts of India. The roots of *C. tora* are used in snake bites and scorpion stings (Chopra *et al.,* 1958). The root extract is effective against shigellosis or bacterial dysentery which is a fatal disease prevalent in Bangladesh (Awal *et al.*, 2004). A paste made from the roots of *C. tora* mixed with lemon juice is applied as a poultice to treat ringworm. The leaf powder is used in the treatment of indigestion and stomach pain and also applied externally in the treatment of skin disease (Manandhar, 2002). Recently, an increasing number of epidemiological and experimental studies have suggested that the consumption of phytoestrogen may have a protective effect on estrogen – related conditions like menopause and estrogen related disease such as prostrate and breast cancers, osteoporosis and cardiovascular diseases (Cos *et al.*, 2003). The ethanolic extract of *C. tora* exhibited a significant stimulation of estrogen dependent MCF – 7 cells, suggesting it's estrogenic activity (EI-Halawany, 2007). The term "Plant tissue culture" is generally used for the aseptic culture of cells, tissues, organs and their components under defined physical and chemical conditions *in vitro* (Street, 1977). An Austrian Botanist, Professor Gottlieb Haberlandt (1854-1945) was closely associated with the beginnings of plant tissue culture. Tissue culture technique in recent years has made easy propagation of many economically and medicinally important plants. Such techniques are also successfully used for crop improvement. Plant tissue culture has also been used in the production of secondary metabolites in plants. It overcomes the barriers in conventional vegetative propagation and fulfils the demand for large scale cultivation in a short period by rapid mass multiplication. Micropropagation is a sophisticated technique for the rapid multiplication of plant (Ahmed, 2001). Micropropagation is *in vitro* clonal propagation from the shoot apical meristem or from node culture to produce a large number of true-to-type plants within a very short period of time (Jha and Ghosh, 2006). Callus culture is often performed in the dark (the lack of photosynthetic capability being no drawback) as light can encourage differentiation of the callus.

The Objectives of the Present Study

☆ To standardize an effective, reproducible and simple protocol With maximum viability for *in vitro* micropropagation from axillary bud explants of *Cassia tora* and

☆ To standardize a protocol for callus induction from leaf, petiole and internode explants of *Cassia tora*.

2. Materials and Methods

2.1. Plant Materials

Healthy and elite twigs of *Cassia tora* L. were collected from Erode. The explant sources were planted under shade house pots and were maintained by timely watering and used for experiments.

2.2. Culture Medium

The Murashige and Skoog (1962) (MS) nutrient medium is used for the present study and the composition of the medium is given below.

Component	Concentration (mg/l)	Component	Concentration (mg/l)
Macro salts		**Micro salts**	
KNO_3	1900	H_3Bo_3	6.20
NH_4NO_3	1650	$MnSO_4.4H_2O$	22.30
$MgSO_4.7H_2O$	370	$ZnSO_4.7H_2O$	8.60
$CaCl_2.2H_2O$	440	$Na_2MoO_4.2H_2O$	0.25
KH_2PO_4	170	$CuSO_4.5H_2O$	0.025
		$CoCl_2.6H_2O$	0.025
Iron source		KI	0.83
$FeSO_4.7H_2O$	27.85		
Na_2 EDTA	37.25	**Vitamins**	
		Meso-inositol	100
Carbon source		Nicotinic acid	0.5
Sucrose	30 g/l	Pyridoxine HCl	0.5
Gelling agent		Thiamine HCl	0.1
Agar	8 g/l	Glycine	4.0
pH of the medium	5.8		

2.3. Preparation of Stock Solution

2.3.1. Macro Salts

Each salt was weighed exactly, dissolved separately to the last particle in a small amount of glass double distilled water; $CaCl_2$ was dissolved and added finally in order to prevent precipitation. Finally all the solutions were pooled together and the volume was made up into 500 ml distilled water.

2.3.2. Micro Salts

Each chemical was weighed exactly, dissolved separately and mixed together. Finally the volume was made up into 500 ml distilled water.

2.3.3. Vitamins

The vitamins were weighed and separately dissolved in distilled water and finally the volume was made up into 250 ml. Meso-inositol was prepared and added freshly.

2.3.4. Iron Source

Each chemical was weighed exactly, dissolved separately, mixed together and made up into 250 ml distilled water. Na_2 EDTA was dissolved in 100 ml distilled water and then gently heated (changed from white colour to light yellow colour). $F_2SO_4. 7H_2O$ was dissolved in 100 ml distilled water and finally made up to 250 ml distilled water.

2.4. Preparation of Stock Solutions of Plant Growth Regulators

Separate stock solutions were prepared for each plant growth regulator by dissolving it in an appropriate solvent (cytokinins in NaOH and auxins in 95 per cent ethanol) and made up into final volume with distilled water. All the stock solutions were stored in brown bottles under refrigeration.

2.5. Preparation of Solid Medium

Appropriate quantities of the various stock solutions, meso-inositol and sucrose were added. The final volume was made with the addition of distilled water. After thorough mixing, the pH of the medium was adjusted to 5.8 (using 0.1 N sodium hydroxide or 0.1N hydrochloric acid) and agar (0.8 per cent w/v; bacteriological grade, Hi media, India) was added into the medium. 10-15 ml of molten agar medium was dispensed into the culture tubes (Borosil, India) (25 x 150 mm) and plugged tightly with non-absorbent cotton. The culture tubes were autoclaved at 15 lb/in² at 121°C for 15 min.

2.6. Explants

Shoot tip, node, leaf and internode were used as explants for the present study.

2.7. Surface Sterilization of Explant

Two types of sterilization methods were followed for this study.

2.7.1. Method 1

Explants like shoot tip, node, internode, petiole and leaf were washed thoroughly in water for 10 min. and placed in 1 per cent (v/v) detergent solution (Teepol/Reckitt Benckiser, India) for 5 min. Then it was placed under running tap water for 30 min. They were then surface sterilized with 70 per cent alcohol for 1-2 min. and 1 per cent (w/v) HgCl₂ solution for 4 to 5 min., followed by repeated washes with sterile distilled water (four to five times, 3 min. each).

2.7.2. Method 2

Explants were washed thoroughly under running tap water for 30 min. and then they were surface sterilized with 70 per cent alcohol for 30 sec and 0.1 per cent (w/v) HgCl₂ solution containing 1ml of 1 per cent detergent (Teepol/Reckitt Benckiser, India) solution for 4 to 5 min, followed by repeated washes with sterile distilled water for five times, 3 min. each.

2.8. Inoculation

Inoculation was carried out in the laminar air-flow chamber (ATLANTIS CLEAN AIR). Before inoculation, the working place was thoroughly cleaned with ethanol and sterilized by UV radiation for 25 min. The sterilized explants were inoculated on the culture medium. The sterile equipments like forceps and blade with holder were sterilized by dipping in 95 per cent alcohol followed by flaming and cooling. Before and after inoculation the mouth of culture tubes were heated over flame. Before starting the inoculation, hands were surface sterilized with 95 per cent alcohol and inoculation was carried out in the vicinity of the flame. Each treatment had 10 replicates.

2.9. Culture Conditions

Cultures were labelled carefully and were kept in a culture room under maintained conditions of temperature (25±2°C) and 16 hrs. of photoperiod with light intensity of 40 µmol/m²/s.

2.10. Micropropagation

The shoot tip of 5-8 mm in length and axilliary bud (node) of 1 cm in length was used for inoculation. Sterilized explants were placed vertically on the MS medium. MS medium supplemented with cytokinins–BA (Benzyl Adenine) and 2- iP (N6 – (2-isopentenyl) adenine) were used to initiate shoot bud sprouting and multiple shoot induction in various concentrations (0.5 – 10.0 mg/l) and for rhizogenesis MS medium was supplemented with auxin NAA (a-Naphthalene Acetic Acid) in various concentrations (0.2 – 1.0 mg/l).

2.11. Callus Induction

Leaf, internode and petiole explants were cultured on MS basal medium containing 35g/l sucrose. The leaf explants were cut into 1cm pieces which included the midrib. The leaf pieces were placed abaxial surface down on the medium. All explants were placed horizontally on the medium.

Media used for callus induction were as follows: MS basal, MS supplemented with various auxins – NAA and 2, 4-D (2-4-Dichlorophenoxy Acetic Acid) in combination with cytokinin (BA). Callus induction was observed from 10-15 days. Cultures were kept in darkness for callus induction.

2.12. Effect of Media and Carbon Source on Culture Media

2.12.1. Media

Different media, *viz.*, MS (Murashige and Skoog, 1962) and WPM – Woody Plant Medium (McCown and Lloyd, 1981) were tested to observe the influence on shoot bud initiation and multiplication.

2.12.2. Carbohydrate

An exogenous supply of carbohydrate for its requirement, of which the plant normally fixes from the atmosphere by photosynthesis, was made to *in vitro* plants through medium. For this purpose three different sugars (fructose, glucose and sucrose) were tested for the multiple shoot induction for shoot tip and axilliary bud explants.

2.13. Statistical Analysis

All experiments were repeated thrice. Each treatment consists of 10 replicates and the data recorded were subjected to statistical analysis according to Duncan's Multiple Range Test (DMRT).

3. Result and Discussion

Cassia tora L. is a well known oriental herb in traditional medicine (Anonymous, 1950). It is a rainy season weed in India. The prime part of the present study was the preparation of contamination free explants. This was achieved by surface sterilization of explants. The following are the details of surface sterilization of explants (Table 8.1).

Table 8.1: Surface sterilization of explants of *Cassia tora*.

Method	Sterilants	Duration of Exposure	Explant	Per cent of Contamination Controlled up to 30 Days
1	Tap water ↓	10 minutes	Shoot tip	81
	1 per cent Teepol ↓	5 minutes	Node, Leaf	
	Tap water ↓	30 minutes	Internode and Petiole	
	70 per cent OHSDW ↓	1 to 2 minutes		
	0.1 per cent HgCl$_2$ ↓	1 time		
	SDW	4–5 minutes 4-5 times	Shoot tip Node, Leaf	87
II	Tap water ↓	30 minutes	Internode and Petiole	
	70 per cent OH ↓	30–45 seconds		
	SDW ↓	1 time		
	0.1 per cent HgCl$_2$ + 1 ml from 1 per cent Teepol ↓	3 minutes 3 ½–4 minutes 5 times		
	SDW			

Surface sterilant 70 per cent alcohol for 30 seconds and 0.1 per cent HgCl$_2$ containing 1 ml of 1 per cent teepol solution followed by 5 rinses with sterile distilled water had given the best sterilization results with least percentage (13 per cent) of contamination.

3.1. Carbon Source

Among three (fructose, glucose and sucrose) sugars tested as carbon source, sucrose showed the best response of multiple shoot tip and axillary bud explants. Glucose and fructose were inefficient.

Full strength MS medium with 3 per cent of sucrose had showed optimum response for shoot tip explants. But for the axillary bud explants, full strength MS medium with 1.5 per cent of sucrose was enough. Sucrose level above 3 per cent produced basal callus in both explants within 5 days.

3.2. Explant

Axillary bud explants showed better response (0.5, 1.0, 2.0 mg/IBA) on MS medium and developed multiple shoots compared to shoot tip explants (Table 8.2). The performance of axillary bud was good in the production of more number of shoots with less callus formation.

Table 8.2: Shoot bud induction and shoot multiplication from shoot bud and axillary bud explants of *Cassia tora* on MS medium supplemented with BA, after 45 days.

Explant	Growth Regulator (mg/l)	Per cent of Response	No. of Shoots/ Explant
Shoot tip	0.5	54.2^g	1.3^{fg}
	1	57.1^f	2.2^c
	2	46.2^{gh}	1.6^{*d}
	4	46.2^{gh}	*1.2^{gh}
	6	28.4^i	*1.2^{gh}
	8	BC	NR
	10	BC	NR
	0.5	100^a	1.6^d
Axillary bud	1	100^a	3.4^a
	2	85.6^b	2.7^b
	4	71.2^c	1.5^e
	6	67.7^{cd}	1.4^{ef}
	8	64.2^{de}	1.3^{fg}
	10	BC	NR

*: Induction of basal callus; BC: Basal Callus; NR: No Response.

Values are mean 10 explants per treatment and repeated three times. Mean Values within a column followed by different letters are significantly different from each other at 5 per cent level comparison by DMRT.

3.3. Micropropagation

Shoot tip and axillary bud explant cultured on MS basal medium without any growth regulator thrived well. The shoot tip explants were cultured on MS basal medium without any growth regulator produced single shoot (4-5 cm) with complete root system. When axillary buds cultured on full strength MS medium without growth regulator produced little basal callus with single shoot but no roots. But the same explant cultured on full strength MS medium with reduced amount of sucrose (1.5 per cent) produced normal shoot and roots without basal callus.

To initiate the study, nodal explant of *Cassia tora* needs surface sterilization and initiation of explants on MS medium. The initiated explants within few days turned black and struggled to survive. It was found that duration of treatment for $HgCl_2$ is very critical due to soft and woody nature of explants. During surface sterilization treatment, it was found that treatment with $HgCl_2$ leads to blackening of the explants. Proper surface sterilization of explants is essential in micropropagation (Yang *et al.*, 1995). Hence the surface sterilization procedure was optimized and this helped in preventing blackening of tissues and establishment of clean cultures. Generally sucrose level enhanced bud break, shoot number and shoot length in the culture of *C.tora*. Very low level of sucrose (2 per cent) declined shoot proliferation, leaves turned light green/yellow/glossy in appearance and shoot became slender.

MS medium supplemented with 1.0 mg/l BA showed maximum response (100 per cent) more number (3.2) of shoots with complete root system (Tables 8.3 and 8.4).Culture medium containing 10.0 mg/l of BA induced basal callus and resulted in more number of shoots (2.8) but the length of the shoot was very less (0.5 cm) (Table 8.3). Culture medium containing 0.5 mg/l of 2-ip produced more number of shoots (2.6) and the shoot length was also maximum (7.8 cm). Ms medium with 10.0 mg/l of 2-ip showed minimum number of shoots (1.1) and the shoot length was also minimum (2.1 cm) (Table 8.3).

The fully grown excised shoots cultured on MS medium supplemented with BA (1mg/l) for rhizogenesis showed 100 per cent shoot bud initiation and BA with 4 mg/l produced maximum root length of 6.3 cm. 8 mg/l and 10 mg/l of BA in MS medium for rhizogenesis produced lower amount of basal callus with nil root initiation (Table 8.4).

The excised shoots cultured on MS medium supplemented with 2-ip (0.5 mg/l) for rhizogenesis recorded maximum (75 per cent) shoot bud initiation with the highest root length of 8.1cm.The increasing concentration of growth regulator (2,4 and 6 mg/l) indicated the decreasing shoot bud initiation and the length of root. The maximum concentration of 2-ip showed high amount of basal callus with no rhizogenesis (Table 8.4).The present observation was found that nodal (axillary bud) explants showed the maximum results compared with shoot tip. Similar result is reported by Rahman *et al.* (1993) in *Caesalpinia pulcherrima*. Joshi *et al.* (2003) in *Foeniculum vulgare* and Anis (2007) in *Cassia angustifolia*.

In contrast, apical bud shows maximum response in pea (Griga *et al.*, 1986), *Cnidium officinale* (Pant *et al.*, 1996), *Pterocarpus santalinus* (Arockiasamy *et al.*, 2000), *Cucumis sativus* (Vasudevan *et al.*, 2001) and *Cardiospermum halicacabum* (Jayaseelan, 2001).

Multiple shoot formation was observed from shoot tip and axillary bud in our study. These results corroborate with previous studies of other plants *viz.*, *Ixora coccinea*

(Lakshmanan *et al.*, 1997), *Ocimum sanctum* (Begum *et al.*, 2000), *Wedelia calendulacea* (Emmanuel *et al.*, 2000) and *Psoralea carylifolia* (Anis and Faisal, 2005).

Table 8.3: Shoot bud induction and shoot multiplication from shoot bud and axillary bud explants of *Cassia tora* on MS medium supplemented with BA and 2ip after 45 days.

Growth Regulator (mg/l)	Shoot Bud Initiation	Bud Initiation (day)	No. of Shoots/ Explants	Shoot Length (cm)
BA				
0.5	78.5[de]	6	1.1[j]	1.8[hi]
1	100[a]	4	3.2[a]	8.3[a]
2	71.3[f]	6	1.6[f]	4.2[cd]
4	64.2[g]	6	1.3[h]	4.1[de]
6	82.1[cd]	7-8	1.2[hi]	1.8[hi]
8	85.5[c]	7-8	*2.8[b]	0.8[j]
10	92.7[b]	7-8	*2.8[b]	0.5[jk]
2 – IP				
0.5	60.6[gh]	10	2.6[c]	7.8[b]
1	57.1[hi]	12	2.4[d]	4.5[c]
2	42.8[j]	12	2.0[e]	4.5[c]
4	28.3[k]	12	1.5[fg]	4.1[de]
6	21.2[l]	12-13	1.3[h]	3.5[f]
8	10.6[m]	13	1.2[hi]	3.2[fg]
10	10.6[m]	13	1.1[jk]	2.1[h]

*: Induction of Basal Callus; NR: No Response.

Values are mean 10 explants per treatment and repeated three times. Mean values within a column followed by different letters are significantly different from each other at 5 per cent level comparison by DMRT.

Ang and Chan (2003) reported that the axillary buds cultured on basal MS medium without any growth regulator produced single shoot with complete root system in *Spilanthes acmella*. This result supports our present study.

In our present investigation, MS medium added with low concentration (1.0 mg/l) of BA produced normal shoots and complete root system without any auxin. The same phenomenon is observed in *Typhonium flagelliforme* (Su *et al.*, 2003) and *Spilanthes acmella* (Ang and Chan, 2003).

Generally, longer shoots are produced at lower BA concentration, whereas more shoots are induced with high BA concentration (Ye *et al.*, 2002). The same findings were obtained in our present study with *C.tora*.

Multiple shoot formation was achieved using MS medium supplemented with 1.0 mg/l BA. Similar finding is reported by Rap and Chopra (1989) in *Cicer arietinum*, Sivakumar and Krishnamoorthy (2000) in *Gloriosa superba*, Alagumanian *et al.* (2004)

in *Solanum trilobatum*, Sivakumar *et al.* (2006) in *Centella asiatica*, Sujatha *et al.* (2007) in *Cicer arietinum*.

Table 8.4: Rhizogenesis of *in vitro* derived shoots of *Cassia tora* on MS medium supplemented with cytokinins, after 45 days.

Growth Regulator (mg/l)	Shoot Bud Initiation	Root Initiation (day)	No. of Roots/ Explants	Root Length (cm)
BA				
0.5	78.5[b]	15	1.5[fg]	3.3[g]
1	100[a]	14	5.1[a]	10.6[a]
2	67.8[cd]	15	3.2[c]	6.2[cd]
4	64.2[de]	16	3.7[b]	6.3[c]
6	60.6[ef]	17	2.1[e]	5.4[e]
8	BC+	NR	NR	NR
10	BC+	NR	NR	NR
2-ip				
0.5	75.0[bc]	20	2.8[d]	8.1[b]
1	57.1[fg]	22-24	1.7[f]	6.3[c]
2	46.3[h]	22-24	1.4[gh]	5.2[ef]
4	46.3[h]	22-24	1.3[hi]	3.3[g]
6	28.5[i]	26-28	1.2[i]	3.1[gh]
8	BC++	NR	NR	NR
10	BC++	NR	NR	NR

BC+: Lower amount of Basal callus; BC++: High amount of Basal Callus; NR: No response.

Values are mean 10 explants per treatment and repeated three times. Mean Values within a column followed by different letters are significantly different from each other at 5 per cent level comparison by DMRT.

In our study, higher concentration of BA induced basal callus in shoot tip and axillary bud explants and this study is supported by Eellarova and Kimakova (1999) in *Hypericum perforatum*.

Higher concentration of NAA induce basal callus during rhizogenesis in *Heliotropium indicum* (Senthil Kumar and Rao, 2007). The same phenomenon was observed for *C.tora* and its rhizogenesis.

George and Sherrington (1984) report that basal callus turned brown and reduced shoot proliferation in many cases. The same observations were found in our study also.

The fully grown excised shoots were cultured on MS medium supplemented with NAA (0.2 mg/l) for rhizogenesis induced maximum number of roots (1.8) with the induction of basal callus. High concentration of NAA in the callus medium (0.6, 0.8 and 1.0 mg/l) produced basal callus without any roots (Table 8.5).

Table 8.5: Rhizogenesis of *in vitro* derived shoots of *Cassia tora* on MS medium supplemented with NAA, after days.

Growth Regulator (mg/l)	Per cent of Shoots Rooted	No. of Shoots Rooted	Root Length (cm)
NAA			
0.2	85.6a	1.8a	10.6a
0.4	42.8b	*1.2b	5.5b
0.6	BC+	NR	NR
0.8	BC+	NR	NR
1	BC+	NR	NR

*: Induction of Basal Callus; BC+: Low amount of Basal Callus; BC++: High amount of Basal Callus; NR: No Response.

Values are mean 10 explants per treatment and repeated three times. Mean Values within a column followed by different letters are significantly different from each other at 5 per cent level comparison by DMRT.

3.4. Callus Induction

Leaf, petiole and internode were cultured on MS medium with different concentrations of 2, 4-D, NAA alone and NAA with BA for callusing. The best result (100 per cent) in terms of percentage response and callus induction was obtained from leaf explant on NAA (1.0 mg/l) + BA (0.5 mg/l) after 7 to 15 days (Tables 8.6 and 8.7).

Table 8.6. Callus induction from *Cassia tora* on MS medium supplemented with 2, 4, D, after 30 days.

Growth Regulator (mg/l)	Callus Induction (per cent)			Morphology of Callus	
	Leaf	Petiole	Internode	Leaf	Petiole and Internode
0	NR	NR	NR	NR	NR
0.5	75	85	66.5	Yellowish brown, friable callus	Greenish white, friable callus
1	100	65	50	Yellowish brown, friable callus	Yellowish brown, friable callus
3	100	45	40	Yellowish brown, friable callus	Yellowish brown, friable callus
5	85	40	40	Dark brown, friable callus	Greenish white, friable callus

Leaf explants cultured on MS medium supplemented with 2, 4-D produced green and friable callus initially and it became brown after two weeks, whereas petiole and internode explants produced yellowish brown and compact callus.

The leaf and internodes segments cultured on MS medium supplemented with NAA and BA were showing small, stiff, root like structure on the upper surface of the explant after 20 days and from its lower surface callus was developed which was yellowish brown and compact. Some roots were thick, small and stiff. Root like structures appeared more on leaf explants compared to internode explants. When auxin (NAA) alone was supplemented to the medium, white, friable callus was developed. In the present study the leaf explants cultured on MS medium supplemented with 2,4-D produced green friable callus. This was supported by the findings of Das *et al.* (2000) in sweet orange, Shawkat Ali and Bushra Mirza (2006) in rough lemon, Agrawal and Sardar (2006) in *Cassia angustifolia*, Shrivastava and Dubey (2007) in *Withania somnifera* and Chao Yanjie in tobacco.

Table 8.7: Callus induction from *Cassia tora* on MS medium.

Growth Regulator (mg/l)		Callus Induction (per cent)		Morphology of Callus	
NAA	BA	Leaf	Internode	Leaf	Internode
0	0	NR	NR	NR	NR
0.5	0	85	45	Yellowish brown, friable callus	White, friable callus
1	0	100	85	Yellowish brown, friable callus	White, friable callus
3	0	85	75	Yellowish brown, friable callus	White, friable callus
5	0	85	75	Yellowish brown, friable callus	Dark brown, friable callus
0.5	0.5	85	70	Brown, compact callus	Brown, compact callus
1	0.5	100	85	Light brown, compact callus	Light brown, compact callus
3	0.5	85	65	Greenish brown, friable callus	Greenish brown, friable callus
5	0.5	85	65	Greenish brown, friable callus	Greenish brown, friable callus

The *in vitro* propagation method is an effective tool for the concentration of medicinal plants. *In vitro* cell and tissue culture technology is envisaged as a means for germplasm conservation to ensure the survival of valuable medicinal plant species, rapid mass propagation for large scale revegetation and for the genetic manipulation. The protocol developed for the *in vitro* micropropagation of the threatened medicinal plant *Cassia tora* may help in the conservation of the species, commercial cultivation and genetic improvement and for the establishment of a large number of uniform pathogen free plants.

The scientific research on *Cassia tora* suggests a huge biological potential of this plant. There is a demand to standardize the toxic properties of *Cassia tora* and their detailed clinical trials. After proper processing, identification and removal of the harmful properties of leaves, they may be utilized to prepare a good, Ayurvedic Formulations and preparations.

The protocol developed for the micropropagation of the promising medicinal plant, *Cassia tora* could be used in the establishment of a large number of uniform pathogen free plants and for germplasm conservation, commercial cultivation, genetic improvement and also for secondary metabolic production.

4. Conclusion

The protocol developed by this tissue culture technique can be useful for the propagation and also for the conservation of the germplasm of this medicinally important threatened plant which can enhance the rate of multiplication and can reduce the time period. Valuable bioresources of medicinal plants are being lost due to lack of awareness, destructive harvesting and unscientific collection.

References

Ahmed, Z., Akhter, F., Haque, M.S., Hasima Banu, Rahman, M.M. and Frauzzaman, A.K.M., 2001. Novel Micropropagation system. *J. Bio. Sci.*, 1(11): 1106–1111.

Awal, M.A., Hossain, M.S., Rahman, M.M., Pravin, S., Bari, M.A. and Hasque, M.E., 2004. Ant shigellosis activity of the root extracts of *Cassia tora* Linn. *Pak. J. Biotech. Sci.*, 7(4): 577–579.

EI–Halaway, A.M., Chung, M.H., Nakamura, N., Ma. C.M., Nishihara, T. and Hattori, M., 2007. Estrogen and anti-estrogenic activities of *Cassia tora* phenolic constituents. *Chem. Pharm. Bull.*, 55(10): 1476–1482.

Jha and Ghosh, B., 2006. *Plant Tissue Culture: Basic and Applied*. Universities Press (India) Pvt. Ltd.

Prajabat, N.D. and Kumar, U., 2003. *Agros Dictionary of Medicinal Plants*. Agrobios Publication, India.

Senthil Kumar, M. and Rao, M.V., 2007. *In vitro* micropopagation of *Heliotropium indicum* Linn.: An Ayurvedic herb. *Indian J. Biotech.*, 6: 245–249.

Scientific Basis of Herbal Medicine (2013)
Editor: **Dr. Parimelazhagan Thangaraj**
Published by: **DAYA PUBLISHING HOUSE, NEW DELHI**

Pages 87–93

Chapter 9

Screening of Edible Mushrooms for L-Ergothionine and Antioxidant Activity

R. Muthu Dharani, D. Ramya Shree, G.J. Sree Meenaskhi, N. Saraswathy and P. Ramalingam

Department of Biotechnology,
Kumaraguru College of Technology,
Coimbatore – 641 049, Tamil Nadu, India

1. Introduction

Edible mushrooms are consumed by humans due to their delicious taste. Besides being tasty, they also have various valuable health benefits like lowering the cholesterol levels, prevention of breast and prostate cancer, weight loss, etc. They have very little fat and digestible carbohydrates and high protein content. Edible mushrooms also possess many other secondary metabolites such as vitamins, essential amino acids and phenolic compounds that may prevent cellular damage caused by free radicals (Alma *et al.*, 2009; Xu *et al.*, 2009). Apart from common antioxidants, edible mushrooms are known to synthesis high quantity of L-ergothioneine (LE) which is very effective in protecting the body from free radicals.

LE is a highly stable, active, non-toxic, naturally occurring antioxidant not synthesized in human body and should be obtained from dietary sources (Hartman, 1990). It is a sulphur containing amino acid showing many beneficial effects including antioxidant and antimutagenic properties in human. Under *in vitro* conditions it has been shown that LE plays a dual role in both energy regulation and in protecting cells from oxidative damage. It is widely distributed in the biological system especially in erythrocytes but its exact role in human

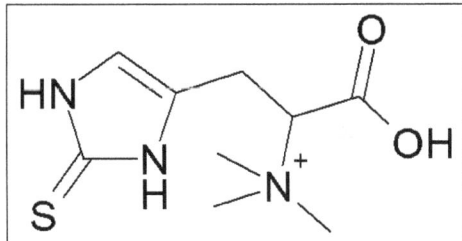

body is not clearly understood. Many basidiomycetes are shown to accumulate LE at different levels in their fruiting bodies and mycelia (Lee *et al.*, 2009). Therefore, the present study was conducted to investigate the LE content and antioxidant activities in the fruiting bodies of the commonly consumed edible mushrooms in Tamil Nadu such as *Agarigus bisporous* and *Pleurotus florida*.

2. Materials and Methods

2.1. Collection of Mushrooms

Fresh fruiting bodies of edible mushrooms; *A. bisporus* (Button mushroom) and *P. florida* (Oyster mushroom) were collected from local markets in Coimbatore.

2.2. Processing of Mushroom Fruiting Body

Fruiting bodies were washed with distilled water to remove the debris, blot dried and cut into small pieces. These pieces were then dried in an oven at 70°C for about 24 hours or until constant dry weight was obtained. The dried fruiting bodies were powdered and sieved through a sieve (No. 60-Mesh size/250 microns sieve opening). The fine powder collected after sieving was packed and stored in an air tight container until further use. Fruiting bodies of both *A. bisporous* and *P. florida* are cooked separately in a stainless steel vessel for 20 min without adding additional water. Cooked fruiting bodies were dried, powdered and stored for further use.

2.3. Preparation of Mushroom Extracts

2.3.1. Hot Water Extract

Hot water extract was prepared by adding 10 g of fruiting body powder prepared earlier from *A. bisporous* and *P. florida* separately to 100 ml of boiling water for 4 hours. The mixture was then filtered through Whattman No. 1 filter paper and the residue was further extracted twice with additional 100 ml of boiling water. The filtrates thus obtained were pooled and kept for drying. The dried extracts were collected in vials and stored at 4°C until further use.

2.3.2. Cold Water Extract

Cold water extract was prepared by adding 10 g of fruiting body of *A. bisporus* and *P. florida* powder separately to 100 ml distilled water (25°C) and subjected to stirring at 100 rpm at 25°C for 24 hours. The resultant mixture was centrifuged at 5000 g for 15 min and filtered through Whattman No. 1 filter paper. The residue was then extracted twice with 100 ml o f distilled water. The filtrates were pooled and kept for drying. The dried extracts were collected in separate vials and stored at 4°C (Paul *et al.*, 2010).

2.3.3. Methanol Extract

Methanol extract was prepared by adding 10 g of fruiting body of *A. bisporous* and *P. florida* powder to 100 ml methanol (25°C) and subjected to stirring at 100 rpm at 25°C for 24 hours. The mixture was then filtered through Whattman No. 1 filter paper and the residue was extracted with two additional 100 ml portions of methanol, as described above. The filtrates collected were pooled and kept for drying. The dried extracts were collected in vials and stored at 4°C (Paul *et al.*, 2010).

2.4. Determination of Total Antioxidant Capacity of Mushroom Extracts

About 0.2 ml of fruiting body extract was added to two ml of phosphor molybdenum solution (0.6 M sulfuric acid, 28mM sodium phosphate and 4 mM ammonium molybdate) and were incubated at 95°C for 90 minutes. The tubes were cooled to room temperature and the absorbance was measured at 695 nm using UV/VIS Spectrophotometer (ELICO SL 149). Total antioxidant activity was calculated by following formula and expressed as ascorbic acid equivalent.

Ascorbic acid equivalent (μM/g) = (T/S)*C*(V/P)*(RS/E)*(1*MW)

- T: OD of test solution.
- S: OD of standard.
- C: Concentration of test (μg).
- V: Volume of solvent used for extraction (ml).
- P: Amount of powder (g).
- RS: Volume of reagent solution (ml).
- E: Volume of extract (ml).
- MW: Molecular weight of ascorbic acid (176-13 g/g mol).

2.5. Determination of DPPH Radical Scavenging Activity

DPPH solution (3 ml) was added to 1 ml of samples containing various concentrations of extracts (1-10 mg/ml). The solution in the tubes was stirred thoroughly and incubated in dark room for 30 minutes at room temperature and the absorbance was measured at 517 nm. The test tubes with an equal volume of DPPH in methanol served as negative control. The DPPH radical-scavenging activity of the extracts in terms of percentage inhibition was calculated using the following equation;

DPPH scavenging activity (per cent) =
$$\{1- (Abs_{517} \text{ sample} / Abs_{517} \text{ DPPH solution})\} \times 100$$

2.6. Determination of ABTS Cation Radical Scavenging Activity

ABTS cation radical was generated by mixing 7 mM ABTS solution with ammonium persulfate (2.45 mM). Then tubes were kept in dark room for 12-16 hours to obtain a dark coloured stock solution. The stock solution was diluted with methanol to obtain 0.7 O.D at 745 nm. Various concentrations of the fruiting body extracts (50-250 μg/ml) were prepared in water. About 0.3 ml of the sample solution was added to 3 ml of ABTS working standard and mixed thoroughly in a microcuvette and the absorbance was measured at 745 nm in 3 min intervals. A solution of ABTS working standard and 0.3 ml of methanol was served as the control. Methanol was used as blank. The ABTS cation radical-scavenging activity of mushroom extracts was calculated in terms of percentage using to the following formula:

ABTS cation radical scavenging activity (per cent) = [(Abs control − Abs sample)/ Abs control] x100 per cent

2.7. Determination of L-ergothioneine Content

Solutions of L-ergothioneine of known concentrations were allowed to react with 2-Py-S-S-2-Py at pH approx. 1; the reaction volume was 3.6 ml; 2.5 ml of 2-Py-S-S-2-Py (1.5 mM), 0.6 ml of 1 M-HCl and 0-0.5 m1 of 50 mM sodium phosphate buffer containing 100 μM $CuSO_4$ (pH9.5) were mixed in both sample and reference cells. After that spectrophotometer had been set to zero at 343 nm, 50 mM sodium phosphate buffer containing 100 μM $CuSO_4$ (pH 9.5) (up to 0.5 ml) was added to the reference cell and the same volume of L-ergothioneine solution (to give a final concentration of up to 40μM) was added to the sample cell. An instantaneous increase in absorbance was recorded. The results of the experiments were used to construct the calibration curve. The extracts were also estimated by the same method by dissolving them in distilled water and from the absorbance value obtained the concentrations of LE were found from the standard graph (slightly modified protocol of Carlsson *et al.,* 1974).

3. Results and Discussions

3.1. LE Content in Fruiting Bodies of *A. bisporous* and *P. florida*

LE is not synthesized in human body but it is present high quantity in erythrocytes, eyes etc (Hartman, 1990). Although this molecule was first described in 1958, its biological role in human body is recently elucidated as its participation in antioxidant activity and inflammatory processes. One of the food materials which are rich in LE is edible mushrooms. The LE content was estimated for hot water, cold water and methanol extract of raw and cooked fruiting bodies of *A. bisporous* and *P. florida* and the results are presented in the Figures 9.1 and 9.2. LE content of extracts from *P. florida* was found to be comparatively higher than *A. bisporous*. Among the various extracts, cold water extract showed highest LE content (41.8 mg/g dry weight) followed by methanol extract (39.5 mg/g dry weight) and hot water extract (21.4 mg/ g dry weight) for raw *P. florida*. Cooking of *P. florida* fruiting body reduced the LE content in all extracts. In *A. bisporous*, the LE content estimation of extracts prepared from raw fruiting bodies showed to highest in cold water extract (13.4 mg/g dry weight) followed by hot water extract (10.7 mg/g dry weight) and methanol extract (9.15 mg/g dry weight). Similarly, cooking of the fruiting body reduced the LE content to some extent. Thus, analysis of extracts from fruiting bodies of *A. bisporus* and *P. florida*, shown that that cold water extracts are good sources of LE while the LE content was reduced due to cooking but not completely destroyed.

3.2. Total Antioxidant Capacity of Fruiting Bodies of *A. bisporous* and *P. florida*

Total antioxidant activity of different mushroom extracts from fruiting bodies of *A. bisporous* and *P. florida* were analysed using reduction of ammonium molybdate and the total antioxidant potential is represented as ascorbic acid equivalent. The

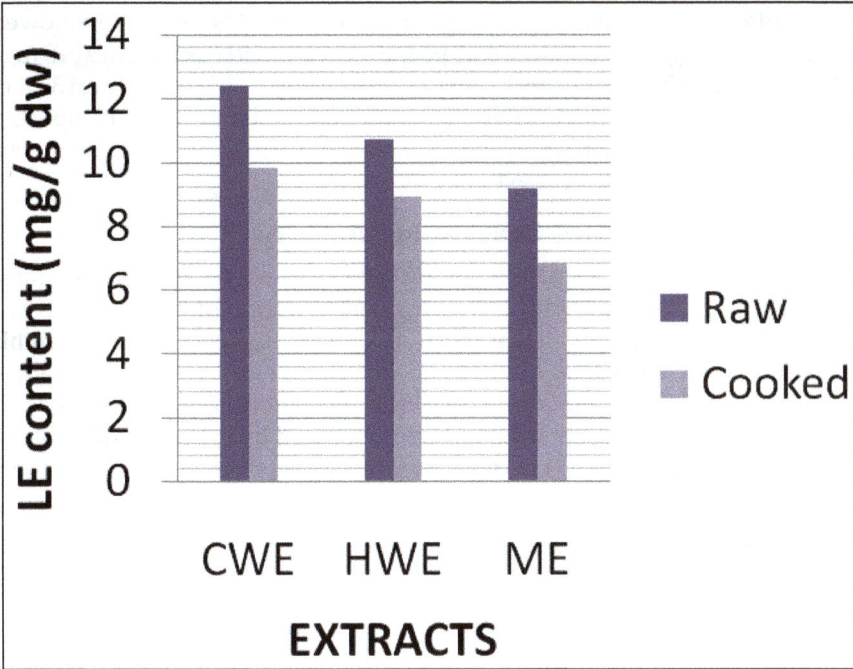

Figure 9.1: LE content in fruiting bodies of *A. bisporus*.

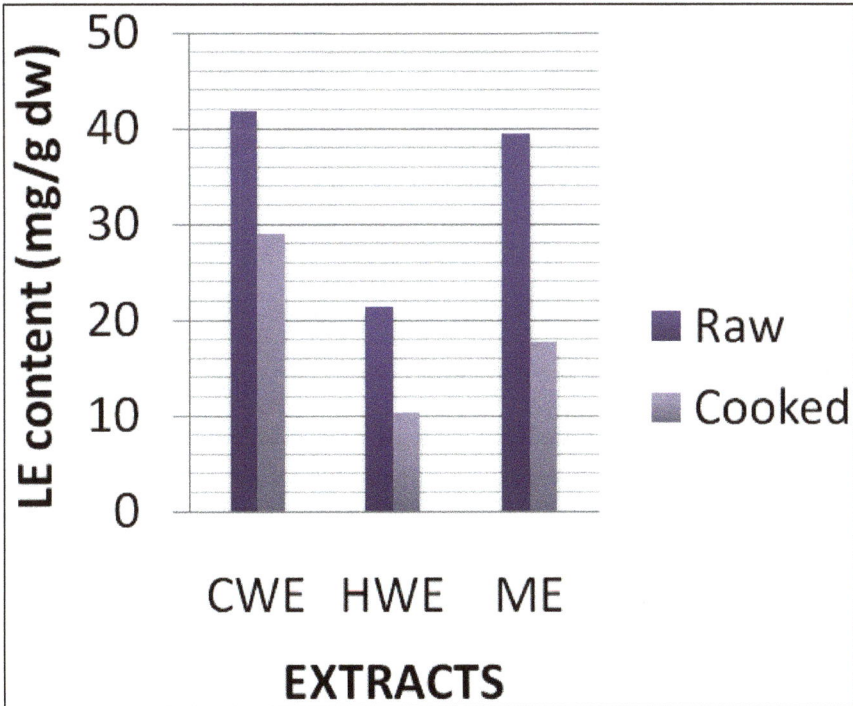

Figure 9.2: LE content in fruiting bodies of *P. florida*.

total antioxidant capacity was highest in cold water extract (94.7µM/g) followed by hot water (92.0 µM/g and methanol 65.0 µM/g ascorbic acid equivalent at 10 mg/ml for *A. bisporous* raw mushroom and it was reduced to 45.5 µM/g, 85.0 and 34.4 µM/g ascorbic acid equivalent cold water, hot water and methanol extract respectively. All the three types of extracts showed gradual increase in total antioxidant activity with increasing concentration of extracts.

3.3. DPPH Free Radical Scavenging Activity

The results of DPPH radical scavenging activity showed that percent inhibition of all extracts was increased with increasing concentration of extracts. The results are presented in Figure 9.3. Cold water extract showed highest percent of inhibition (79.95 per cent) of   free radical followed by hot water extract (75.08 per cent) and methanol extract (58.34 per cent) for *A. bisporous*. The percent inhibition was found to be highest in cold water extract (84.2 per cent) followed by methanol (82.3 per cent) and hot water extract (73.1 per cent) for *P. florida*. Among the two edible mushrooms, tested *P. florida* showed higher radical scavenging capacity than *A. bisporous*.

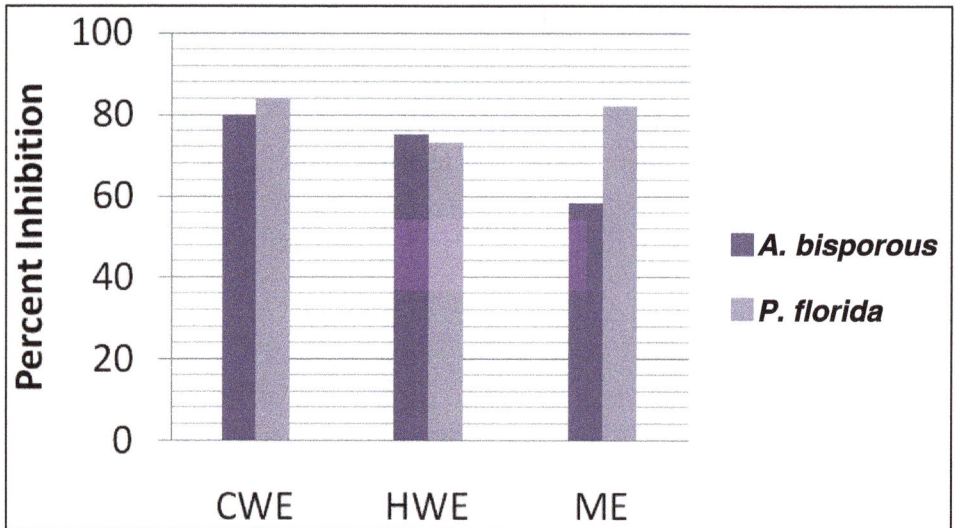

Figure 9.3: DPPH free radical scavenging activity.

3.4. ABTS Cation Radical Scavenging Activity

The results of ABTS radical scavenging activity revealed that percent inhibition of all extracts were increasing with increasing concentration of extracts. The results are presented in Figure 9.4. The percent inhibition was found to be the highest in hot water extract (94.06 per cent) followed by cold water extract (81.66 per cent) and methanol extract (44.01 per cent) for *A. bisporous*. Both hot water and cold water extracts of showed 92 per cent inhibition and methanol extract showed 80 per cent ABTS radical scavenging activity for *P. florida*.

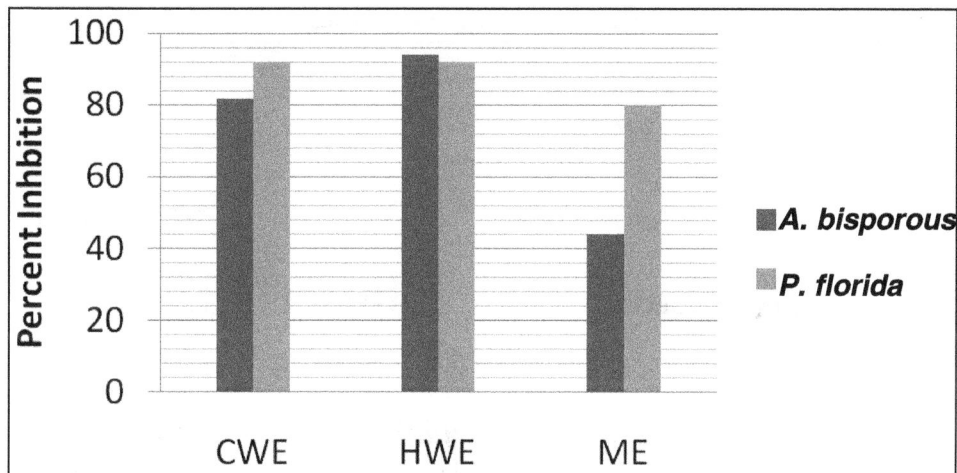

Figure 9.4: ABTS cation radical scavenging activity.

4. Conclusion

The present study has shown that mushroom extracts have potential antioxidant properties in DPPH and ABTS radical scavenging assays. Further, cold water and hot water extracts showed significant antioxidant properties when compared with methanol extracts. And the content of L-ergothioneine was higher in *P. florida* when compared with *A. bisporus*. The comparison of these antioxidant activities with that of the L-ergothioneine contents showed a relation between the EC_{50} values of DPPH radical scavenging effect and the ABTS cation radical scavenging effect with that of LE content.

References

Alma, E., Estrada, R., Lee, H.J., Beelman, R.B., Gasco, M.D. and Royse, D.J., 2009. Enhancement of the antioxidants ergothioneine and selenium in *Pleurotus eryngii* var. *eryngii basidio mata* through cultural practices. *World J. Microbiol. Biotechnol.*, 25: 1597–1607.

Carlsson, J., Kibrsatn, M.P. and Brocklehurst, K., 1974. A Convenient Spectrophotometric Assay for the Determination of L–Ergothioneine in Blood. *Biochem. J.*, 139: 237–242.

Hartman, P.E., 1990. L-Ergothioneine as antioxidant. *Meth. Enzy.*, 186: 310–318.

Lee, W.J., Park, E.J., Ahn, J.K. and Ka, K.H., 2009. Ergothioneine contents in fruiting bodies and their enhancement in mycelial cultures by the addition of methionine. *Mycobiology*, 37: 43–47.

Paul, B.D. and Snyder, S.H., 2010 The unusual amino acid L-ergothioneine is a physiologic cytoprotectant. *Cell Death Differ*, 17: 1134–1140.

Xu, W., Zhang, F., Luo, Y., Ma, L., Kou, X. and Huang, K., 2009 Antioxidant activity of a water-soluble polysaccharide purified from *Pteridium aquilinum*. *Carbohydrate Research*, 344: 217–222.

Scientific Basis of Herbal Medicine (2013)
Editor: Dr. Parimelazhagan Thangaraj
Published by: DAYA PUBLISHING HOUSE, NEW DELHI

Pages **95–100**

Chapter 10

Phytochemical and Bio-efficacy Studies on *Plumbago rosea* L.

T. Renisheya Joy Jeba Malar and M. Johnson

Centre for Plant Biotechnology,
Department of Plant Biology and Plant Biotechnology,
St. Xavier's College (Autonomous), Palayamkottai, Tamil Nadu, India

1. Introduction

In fact, plants produce a diverse range of bioactive molecules making them a rich source of different types of medicines. Higher plants, as sources of medicinal compounds, have continued to play a dominant role in the maintenance of human health since ancient times (Farombi, 2003). Secondary metabolites produced by plants constitute a source of bioactive substances and nowadays the scientific interest has increased due to the search for new drugs from plant origin (Gothandam *et al.*, 2010). *Plumbago rosea* L. (Family: Plumbaginaceae), a perennial evergreen shrub with about 2 to 4 feet in height is traditionally used in skin disease, anaemia, irregular menstruation and leucorrhoea in the southeast area of Bangladesh (Yusuf *et al.*, 2007). The roots contain an alkaloid called plumbagin, a natural napthaquinone (5-hydroxy-2-methyl-1,4-naphthoquinone), possessing various pharmacological activities such as antimalarial (Didry *et al.*, 1994), antioxidant activity, anticancer, cardiotonic, antifertility action, antibiotic and antineoplastic (Nahak and Sahu, 2011). *Plumbago* species are reported in the literature for its biological activities such as: antiparasitic (Chan-Bacab and Peña-Rodríguez, 2001), insect anti-feedant (Villavicencio and Perez-Escandon, 1992), antitumoral (Devi *et al.*, 1994) and others, some of them attributed to the presence of special chemical compounds, such as naphthoquinones. All parts of *P. rosea* were used, but the roots have fascinated the chemists and biologists due to tremendous pharmacological properties. The pulped roots or aerial parts are reported abortifacient, while powdered bark, root or leaves are used to treat gonorrhoea, syphilis, tuberculosis, rheumatic pain, swellings, and wound healing (Thakur *et al.*, 1989). Root decoction with boiled milk is swallowed to treat inflammation in the mouth, throat and chest. A paste of the root in vinegar, milk, and water is considered significant against influenza and black water fever, while root infusion is taken orally to treat shortness of breath (Teshome *et al.*, 2008). *Plumbago* species has been described for its significant anticancer (Xu *et al.*, 2010), antitumor (Yang *et al.*, 2010), anti-inflammatory (Sivakumar *et al.*, 2005), antimycobacterial (Patil *et al.*, 2011) and antimicrobial activities (Ravikumar *et al.*, 2011). The plant is also effective against rheumatic pain, sprains, scabies, skin diseases, and wounds. The roots of the plant and its constituents are credited with potential therapeutic properties including antiatherogenic, cardiotonic, hepatoprotective, neuroprotective, and central nervous system stimulating properties (Bagla *et al.*, 2011; Siddique *et al.*, 2011). With this background the present study was aimed to reveal the phytochemical and antibacterial activities of *P. rosea* against the selected human pathogens *viz.*, *P. aeroginosa*, *S. aureus*, *P. vulgaris*, *K. pneumoniae* and *B. subtilis*.

2. Materials and Methods

2.1. Collection of Plant Material

The aerial parts of *Plumbago rosea* were collected from the natural habitats at Chankanachari, Kerala, India. The plant materials were washed under running tap water to remove the surface contamimants and the leaves and stem were separated mechanically. The separated parts were air dried under shade. The dried sample was powdered using mechanical grinder and used for further extraction.

2.2. Extraction of Plant Material

25 g of air dried powder of the sample was extracted successively with organic solvents with 150 ml of solvents *viz.*, petroleum ether, chloroform, methanol and water with the increasing order of polarity using soxhlet apparatus. The extraction was carried out for 8 hours and the extract was concentrated by evaporation in a rota-vaccum. The extract obtained was used for the further assessment of phytochemical and antibacterial activity.

2.3. Preliminary Phytochemical Screening

To reveal the presence of steroids, terpenoids, cardiac glycosides, saponins, tannins, phenolics, amino acids, alkaloids, the preliminary phytochemical screening of various extracts of *Plumbago rosea* was carried out according to the method described by Harborne (1998).

2.4. Antibacterial Activity

Screening of antibacterial activity was performed by disc diffusion technique (Mukherjee, 2004). The methanolic extracts with various concentrations (50, 100, 150, 200, 250 µg/ml) were screened for antibacterial studies against selected human pathogens *viz.*, *P. aeroginosa*, *S. aureus*, *P. vulgaris*, *K. pneumoniae* and *B. subtilis* (Hardi *et al.*, 2004). Commercially available antibiotic disc Amikacin was implanted along with the crude extract disc on the surface of the Muller-Hinton agar plates, which is used as a positive control against Gram positive and Gram negative microbe respectively. The inoculated plates were incubated at 37°C for 18-24 hours and the zone of inhibition was measured and the results were tabulated.

3. Results and Discussion

3.1. Preliminary Phytochemical Screening

The secondary metabolites are responsible for the therapeutic properties of plants and the composition of these secondary metabolites varies from plant species to species. The composition of these compounds with the same species of plant can vary with the nutrient composition of the soil, climatic season, development stage of the plant and natural association with other plants (Renisheya Joy Jeba Malar *et al.*, 2012; Arunkumar and Muthuselvam, 2009). In the present study also preliminary phytochemical screening of nine different chemical compounds (steroids, saponins, phenolics, tannin, alkaloids, anthroquinone, cardiac glycosides, amino acids and terpenoids) were tested in four different extracts. Thus out of 36 (9 x 4 = 36) tests for the presence or absence of the above compounds, only 17 gave positive results and the remaining 19 gave negative results. The 17 positive results show the presence of steroids, saponins, tannin, anthroquinone, amino acid, terpenoids, phenolics, cardiac glycosides and alkaloids. Among the four different extracts, methanolic extract of *P. rosea* showed the maximum (7/9) presence *viz.*, steroids, tannin, alkaloids, phenolics, saponin, cardiac glycosides, and terpenoids. Chloroform extract of *P. rosea* showed the presence only five (5/9) compounds *viz.*, tannin, alkaloid, phenolics, cardiac glycosides and steroid. Petroleum ether extract showed the presence of only minimum (3/9) compounds (phenolics, saponin and steroid). Aqueous extract showed the presence of only two (2/9) compounds such as tannin and phenol (Table 10.1). Sheeja *et al.*, 2011 revealed the presence of phytoconstituents with vaired degree in *P. rosea* petroleum ether, chloroform, acetone, ethanol and aqueous extracts. They observed that ethanolic extract of *P. rosea* showed the presence of maximum number of (5) compounds *viz.*, naphthoquinone, carbohydrates, glycosides, tannins, flavanoids and saponins. But in the present study we observed seven metabolites present in the methanolic extracts of *P. rosea viz.*, steroids, tannin, alkaloids, phenolics, saponin, cardiac glycosides, and terpenoids. The results of preliminary phytochemical analysis confirmed the presence of tannin, saponin and cardiac glycosides; in addition

Table 10.1: Phytochemical screening of *P. rosea* various extracts.

Test	Plumbago rosea			
	Methanol	Chloroform	Pet ether	Aqueous
Steroid	+	+	+	−
Alkaloid	+	+	−	−
Phenol	+	+	+	+
Saponin	+	−	+	−
Tannin	+	+	−	+
Anthroquinone	−	−	−	−
Aminoacid	−	−	−	−
Terpenoid	+	−	−	−
Cardiac glycosides	+	+	−	−

we observed the presence of steroids, alkaloids, phenolics and terpenoids in the methanolic extracts of *P. rosea*. Sheeja *et al.*, observed five compounds *viz.*, cardiac glycosides, glycosides, tannins, flavanoids, proteins and saponins presence in the aqueous extracts of *P. rosea*. Contrary to Sheeja *et al.*, observation, in the present study aqueous extract showed the presence of only two (2/9) compounds such as tannin and phenol.

3.2. Antibacterial Activity

The *in vitro* antibacterial activity of methanolic extract of *P. rosea* was assessed by the agar disc diffusion method. The results were compared with the standard antibiotic Amikacin. 250 µg/ml methanolic extract of *P. rosea* showed good activity against *K. pneumoniae* (20 mm), *B. subtilis* (19 mm), *S. aureus* (17 mm) and moderate activity was observed against *P. vulgaris* and *S. aureus* (16 mm) and lowest activity was observed in 50 µg/ml of methanolic extract of *P. rosea* against *P. vulgaris* (6 mm). The antibiotic Amikacin showed highest zone of inhibition against *K. pneumoniae* (12 mm) and lowest activity was observed against *P. vulgaris* with 9 mm (Table 10.2). Haribabu Rao *et al.*, 2012 revealed that the aqueous extracts of *P. zeylanica* showed no inhibitory zones where as butanol extracts of *P. zeylanica* showed best inhibitory activity than benzene extracts. In the present study we observed the best inhibitory activity in the methanolic extracts of *P. rosea*. Shafiqur Rahman and Nural Anwar (2007) reported the antimicrobial activity of the crude ethanolic extract (250 µg/disc and 500 µg/disc) of *P. zeylanica* against the pathogenic bacteria *viz.*, *B. subtilis*, *B. cereus*, *B. megaterium*, *S. aureus*, *E. coli*, *V. cholerae*, *S. sonnei*, *S. typhi*, *S. paratyphi*, *P. mutabilis*) using the disc diffusion method. The ethanolic extract exhibited zone of inhibitions ranged from 8 to 18 mm in diameter with 250 µg/disc and 16-30 mm in diameter with 500 µg/disc concentration against the test bacteria. In the present study also we tested various concentrations of methanolic extracts using disc diffucion method. We also observed highest zone of inhibition 250 µg/ml methanolic extract of *P. rosea* against *K. pneumoniae* (20 mm), *B. subtilis* (19 mm), *S. aureus* (17 mm) and moderate activity was observed against *P. vulgaris* and *S. aureus* (16 mm). The present study results clearly showed that the methanolic extracts of *P. rosea* had significant and considerable antibacterial activity against various pathogens and further evaluation is necessary to find out the active principle compound responsible for bioactivity.

Table 10.2: Antibacterial activity of *Plumbago rosea* Methanolic extracts.

Pathogens	Zone of Inhibition (mm)					
	Concentration of Extracts in µg/ml					
	Amikacin	50	100	150	200	250
P. aeroginosa	10	9	11	13	14	16
S. aureus	11	7	10	12	16	17
P. vulgaris	9	6	8	10	14	16
K. pneumoniae	12	11	13	15	17	20
B. subtilis	11	7	9	12	14	19

Acknowledgement

The authors are thankful to St. Xavier's College management for providing infrastructure, constant support and encouragements. Financial Support: The author (Renisheya Joy Jeba Malar Tharmaraj) is thankful to Department of Science and Technology, Govt. of India for providing financial assistance through DST-INSPIRE Fellowship (Ref. No. IF110640).

References

Arunkumar, S. and Muthuselvam, M., 2009. Analysis of Phytochemical Constituents and AntimicrobialActivities of *Aloe vera* L. against clinical pathogens. *World Journal of Agricultural Sciences*, 5(5): 572–576.

Bagla, V.P., McGaw, L.J. and Eloff, J.N., 2011. The antiviral activity of six South African plants traditionally used against infections in ethnoveterinary medicine. *Vet. Microbiol.*, doidx.doi.org/10.1016/j.vetmic. 09.015.

Chan-Bacab, M.J. and Peña-Rodríguez, L.M., 2001. Plant natural products with leishmanicidal activity. *Nat. Prod. Rep.*, 18: 674–688.

Devi, P.U., Solomon, F.E. and Sharada, A.C., 1994. *In vivo* tumor inhibitory and radiosensitizing effects of an Indian medicinal plant, *Plumbago rosea* on experimental mouse tumors. *Indian J. Exp. Biol.*, 32: 523–528.

Didry, N., Dubrevil, L. and Pinkas, M., 1994. Activity of anthraquinonic and naphthoquinonic compounds on oral bacteria. *Die Pharmazie*, 49: 681–683.

Farombi, E.O., 2003. African indigenous plants with chemotherapeutic potentials and biotechnological approach to the production of bioactive prophylactic agents. *African. J. Biotech.*, 2: 662–671.

Gothandam, K.M., Aishwarya, R. and Karthikeyan, S., 2010. Preliminary screening of antimicrobial properties of few medicinal plants. *Journal of Phytology*, 2: 1–6.

Harborne, J.B., 1998. *Phytochemical Methods: A Guide to Modern Techniques of Plant Analysis*, 3rd edn. Chapman and Hall, New York, pp. 1–150.

Hardi, A.L. and Uddin, M.I., 2004. Seasonal variation in the intestinal bacterial flora of hybrid Tilapia (*Oreochromis niloticus* x *Oreochromis aureus*) culture in earthen pond in Saudi Arabia. *Aquaculture*, 229(124): 37–44.

Haribabu Rao, D., Vijaya, T., Ramana Naidu, B.V., Subramanyam, P. and Jayasimha Rayalu, D., 2012. Phytochemical Screening and antimicrobial studies of compounds isolated from *Plumbago zeylanica*. L *IJAPBS*, pp. 82–90.

Mukherjee, K.L., 2004. *Medical Laboratory Technology*. Tata McGraw Hill Publishing Company Ltd., New Delhi.

Nahak, G. and Sahu, R.K., 2011. Antioxidant activity of *Plumbago zeylanica* and *Plumbago rosea* belonging to family plumbaginaceae. *Natural Product: An Indian Journal*, 7(2): 51–56.

Patil, C.D., Patil S.V., Salunke, B.K. and Salunkhe, R.B., 2011. Bioefficacy of *Plumbago zeylanica* (Plumbaginaceae) and *Cestrum nocturnum* (Solanaceae) plant extracts against *Aedes aegypti* (Diptera: Culicide) and nontarget fish *Poecilia reticulata*. *Parasitol. Res.*, 108: 1253–1263.

Ravikumar, V.R. and Sudha, T., 2011. Phytochemical and Antimicrobial Studies on *Plumbago zeylanica* (L) Plumbaginaceae. *IJRPC*, 1: 185–188.

Renisheya Joy Jeba Malar, T., Johnson M., Nancy Beaulah, S., Laju, R, S., Anupriya, G. and Renola Joy Jeba Ethal, T., 2012, Anti-bacterial and antifungal activity of *aloe vera* gel Extract. *IJBAR*, 3(3).

Shafiqur Rahman and Nural Anwar, 2007. Antimicrobial activity of crude extract obtained from the root of *Plumbago zeylanica*. *Bangladesh J. Microbiol.*, 24(1): 73–75.

Sheeja, M., Joshi, S.B. and Jain, D.C., 2011. Antiovulatory and estrogenic activity of *Plumbago rosea* leaves leaves in female albino rats. *Indian J. Pharmacol.*, 41(6): 273–277.

Siddique, Y.H., Ara, G., Faisa, M. and Afzal, M., 2011. Protective role of *Plumbago zeylanica* extract against the toxic effects of ethinylestradiol in the third instar larvae of transgenic *Drosophila melanogaster* (hsp70–lacZ)Bg9 and cultured human peripheral blood lymphocytes. *Alternative Medicine Studies*, 1: 726–729.

Sivakumar, V., Prakash, R., Murali, M.R., Devaraj, H. and Devaraj, S.N., 2005. *In vivo* micronucleus assay and GST activity in assessing genotoxicity of plumbagin in Swiss albino mice. *Drug. Chem. Toxicol.*, 28: 499–507.

Teshome, K., Gebre-Mariam, T., Asres, K., Perry, F. and Engidawork, E., 2008. Toxicity studies on dermal application of plant extract of *Plumbago zeylanica* used in Ethiopian traditional medicine. *J. Ethnopharmacology*, 117: 236–248.

Thakur, R.S., Puri, H.S. and Husain, A., 1989. *Major Medicinal Plants of India*. Central Institute of Medicinal and Aromatic Plants, Lucknow, India.

Villavicencio, M.A. and Perez-Escandon, B.E., 1992. Plumbagin activity (from *Plumbago pulchella* Boiss. Plumbaginaceae) as a feeding deterrent for three species of Orthoptera. *Folia Entomol Mex.*, 86: 191–198.

Xu, K.H. and Lu, D.P., 2010. Plumbagin induces ROS–mediated apoptosis in human *Promyelocytic leukemia* cells *in vivo*. *Leuk Res.*, 34: 658–65.

Yang, S.J., Chang, S.C., Wen, H.C., Chen, C.Y., Liao, J.F. and Chang, C.H., 2010. Plumbagin activates ERK1/2 and Akt via superoxide, Src and PI3–kinase in 3T3–L1 Cells. *Eur. J. Pharmacol.*, 638: 21–28.

Yusuf, M.A., Wahab, M.Y., Chowdhury, J.U. and Begum, J., 2007. Some tribal medicinal plants of Chittagong Hill Tracts, Bangladesh. *Bangladesh J. Plant Taxon.*, 14: 117–128.

Scientific Basis of Herbal Medicine (2013)
Editor: Dr. Parimelazhagan Thangaraj
Published by: DAYA PUBLISHING HOUSE, NEW DELHI

Pages 101–105

Chapter 11

Conservation of *Maerua arenaria* Hook. f. and Thoms. (Capparidaceae): A Traditional Medicinal Plant Species

B. Arthi Rashmi, A. Latha and P. Sivaselvi

*Department of Bioinformatics,
Sri Krishna Arts and Science College, Coimbatore – 8, T.N.*

1. Introduction

The genus of Maerua belongs to the family Capparaceae which is widely distributed in India, Pakistan, Africa and Saudi Arabia. The family Capparaceae contains isocyanate glycosides which are know to possess anti-thyroid activity. In Karachi the genus Maerua is common in sandy and stony grounds, straggling among bushes in Malir Cantt. The fleshy roots of this plant used as alternatives, tonic and stimulant. The plant is also used for the treatment of snake-bite and scorpion-sting Literature survey revealed that only three compounds are reported from the genus Maerua and no phytochemical investigations on this plant. Here in we report the isolation and structure elucidation of dodecanoic acid, β-sitosterol, ursolic acid, 4-hydroxybenzoicro acid, methyl grevillate, glycerol 1, 3-didodecanote,1-*O*-coumaroylglycerol and β-sitostreol,3-*O*-β-glucopyranoside respectively from the ethylacetate fraction.

2. Materials and Methods

The plant materials are collected. Experimental of this plant the column chromatography is used. Its contain Aluminium sheets with silica gel were used for TLC to check the purity of compounds and and were visualized under the UV light (254 and 366 nm) followed by ceric sulphate as spraying reagent.

The shade dried whole plant of *Maerua arenaria* (10 kg) was extracted with MeOH (3×40L) at room temperature but mostly this plant root extract only collected. The combined methanolic extract was evaporated under the reduce pressure to obtain a thick gummy mass (300 g). It was suspended in water and successively extracted with n-hexane, ethylacetate n-butanol. The ethylacetate soluable fraction (40 g) was subjected to column chromatography over silica gel eluting with n-hexane, n-hexane ethylacetate, ethylacetate, ethylacetate-methonal, and methonal in increasing the order of polarity.The fraction obtained from n-hexane-ethylacetate gave to 2 major spots on TLC, was subjected to flash chromatography using the solvent system n-hexane-ethylacetate as eluent collecting the 90 fractions of 10 ml each. Thus this experiment is proofed.

3. Discussions

3.1. *Maerua arenaria* (Hook and Thomas)

Flowers

Flowers usually in dense, corymbose racemes, greenish-white, pedicellate. They look beautiful with mainly the greenish stamens radiating out. Sepals 4, ovate-elliptic, acute or slightly acuminate.

Fruits

Fruit cylindrical, 3-8 cm long, 1-1.5 cm broad, torulose or irregularly many knotted, pale brown, often somewhat twisted. A very variable species with regards to hairiness and size of leaves.

Figure 11.1: Flowers.

Figure 11.2: Fruits.

Figure 11.3: Roots (Tubers).

Roots (Tubers)

The tap root is very stout, woody, runs deep in the soil for several meters. Irregularly shaped. The fresh roots are soft, 1-9 cm in diameter, with a smooth brownish surface and concentric deep furrows. The dry roots of the commercial drug occur in 2 to 4 cm long pieces very hard. They are dark brown on the outer side and pale yellow on the inner side.The drug has a sweet taste. No characteristic odour.

3.2. Medicinal Uses of *Maerua arenaria*

The tuber is medicinal and is eaten to quench thirst. Traditionally, the fleshy roots of this plant are used as alternative tonic and stimulant. The plant is also used for treatment of snake bite and scorpion sting.

3.3. Cultivation

Large interest groups get this plants collected from the fields and thus they endanger the plant species in natural conditions and many plants have come in the category endangered plants due to their over exploitation. Medicinal Plants play an important role in human life to combat diseases since time immemorial. The rural folks and tribals in India even now depend largely on the surrounding plants/ forests for their day-today needs. Medicinal plants are being looked upon not only as a source of health care but also as a source of income. The value of medicinal plants related trade in India is of the order of 5.5 billion US dollar and is further increasing day-by-day.

The following questions must be examined before undertaking medicinal plant cultivation:

1. Type of soil, its pH, and irrigation water.
2. We want to check whether the medicinal plant should adopt to the local conditions or not.
3. Proper timing for sowing and harvesting.
4. Market potential.
5. Quality control.
6. Government regulations for export.
7. Management practices.
8. The farmers have land, water and money but they do not know which medicinal plants to cultivate for commercial purpose. So, making awareness among farmers about commercial medicinal plants is essential

3.4. Conservation

Medicinal plant conservation strategies need to be understood and planned for based on an understanding of indigenous knowledge and practices Many drugs contain herbal ingredients, and it has been said that 70–80 per cent of the world's population relies on some form of non-conventional medicine and around 25–40 per cent of all prescription drugs contain active ingredients derived from plants in the United States alone. Many countries rely on these medicinal plants for the health and well being of its population. Conservation is an ethnic of resource use, allocation, and protection. Medicinal plant conservation strategies need to be understood and planned for based on an understanding of indigenous knowledge and practices. Plants have the ability to synthesize a wide variety of chemical compounds that are used to perform important biological functions. As it is endemic and highly demanded one, proper conservation strategies are required through modern propagation techniques.

The most serious proximate threats when extracting medicinal plants generally are habitat loss, habitat degradation, and over harvesting.

4. Conclusion

The study of this plant species have high amount of the secondary metabolites. These compounds are very essential in pharmaceuticals. This species is threatened due to anthropogenic activities and also going to be extinct. So, there is need to conserve the traditional species by the new modern techniques *i.e., in vitro* regeneration, vegetative propagation etc.

5. References

2008. Phytochemical studies of *Maerua Arenaria. J. Chem. Soc. Pak,* 30(1).

Chopra, A.K., 2007. *Medicinal Plants: Conservation, Cultivation and Utilization,* p. 411.

Gupta, Rajini. *Plant Taxonomy: Past, Present and Future,* p. 78.

Khan, I. and Ajami, Dari, 1999. *Global Biodiversity Conservation Measures.*

Khare, C.P., 2007. *Indian Medicinal Plants: An Illustrated Dictionary,* p. 796.

Panda, H., 2004. *Handbook of Herbal Medicines,* p. 334.

Pandey, H.N., Barik, S.K. and Tripathi, O.P., 2006. *Ecolgy, Diversity, and Conservation of Plants and Ecosystems in India,* p. 264.

Sahoo, S., 2001. *Conservation and Utilization of Medicinal and Aromatics Plants,* p. 67.

Sharma, O. P., 1993. *Plant Taxonomy,* p. 207.

Vardhana, R., 2008. *Direct Uses of Medicinal and their Identification,* p. 216.

Scientific Basis of Herbal Medicine (2013)
Editor: **Dr. Parimelazhagan Thangaraj**
Published by: DAYA PUBLISHING HOUSE, NEW DELHI

Pages 107–114

Chapter 12

Phytochemical and Antioxidant Activity of Crude Alkaloid Extract of *Hybanthus Enneaspermus* (L.)

S.K. Reshmi, M. Kathiresh, R. Rakkimuthu,
K.M. Aravinthan

Research Department of Biotechnology,
Dr. Mahalingam Center for Research and Development,
NGM College, Pollachi – 642 001, India

1. Introduction

Medicinal plants are of great importance to the health of individuals and communities. The medicinal value of these plants lies in some chemical substances that produce a definite physiological action on the human body and these chemical substances are called phytochemicals (Subhashini *et al.*, 2010). These are non-nutritive chemicals that have protective or disease preventive property. The most important of these phytochemicals are alkaloids, flavonoids, tannins and phenolic compounds (Hill, 1952). Alkaloid is considered to be the major class of secondary plant substance and its found to possess various factors such as antibacterial, antiviral, antihypertensive, antitumour, antiarrhythmic and antioxidant activity. Oxidative damage in the human body plays an important causative role in disease initiation and progression. Antioxidants is a compound that can significantly delay or prevent the oxidation of substrate even if the compound is present in a significantly lower concentration than the oxidized substrate and can be recycled in the cells or irreversibly damaged (Halliwell and Gutteridge, 2007). There are several reports in the literature regarding the anti-oxidant activity of crude extracts prepared from plants (Maryam *et al.*, 2009; Sasikumar *et al.*, 2010; Mandal, *et al.*, 2010). As a result, some natural products have been approved as new antioxidant drugs, but there is still an urgent need to identify novel substances.

Hybanthus enneaspermus is a herbal plant used for medicinal purpose. It is also called as "hump back flower" and they are also called as green violet (L) *F. muell*. It belongs to violaceae family. It is perennial often creeping and widely distributed in the tropical and sub tropical regions in the world. It grows 15–30 cm in height with ascending nature (Ibrahim *et al.*, 2008). The leaves and tender stalks are demulcent and used as a decoction or electuary; mixed with oil, they are employed in preparing a cooling ointments for the headache. The root is diuretic and administered as an infusion in gonorrhea and urinary infections. They are very effective in the treatment of infectious disease. The plant possesses anti-inflammatory, antiplasmoidal, antimicrobial (Rajakaruna *et al.*, 2002; Weniger *et al.*, 2004), anti-convulsant (Kirtikar and Basu, 1991; Hemalatha *et al.*, 2003) and also used to treat diarrhea, dysuria, urinary tract infections, male sterility and diabetes because which possess many bioactive components such as phenol, alkaloids and flavanoids (Yoganarasimhan, 2000; Patel *et al.*, 2011). So the present study was aimed to investigate the phytochemical and antioxidant activities of *Hybanthus enneaspermus* leaves.

2. Materials and Methods

2.1. Sample Collection

Hybanthus enneaspermus leaves were collected from Coimbatore, Tamil Nadu, India and stored in sealed polyethylene bags at -20°C until extraction.

2.2. Solvent Extraction

The powdered plant materials (10 gm) were extracted with 100 ml of methanol in a shaker for 72 hrs. The extract was concentrated to remove the solvent and filtered through whattman No. 1 filter paper. The clear extract was used for preliminary screening for bio-active compounds.

2.3. Preliminary Screening

2.3.1. Qualitative Phytochemical Analysis

Chemical tests were carried out on the methanolic leaf extract for the qualitative determination of phytochemical constituents as described by Harborne (1973), Trease and Evans (1989) and Sofowara (1993).

2.3.2. Alkaloids

1 ml of methanol extract was divided into three portions. Mayer's reagent was added to one portion and Draggendoff's reagent to the other and Wagner's reagent to the other portion. The formation of a cream (with Mayer's reagent) or reddish brown precipitate (with Draggendoff's reagent) and yellow to orange precipitate (with Wagner's reagent) was regarded as positive for the presence of alkaloids.

2.3.3. Flavonoids

5 ml of dilute ammonia solution were added to a portion of the methonal filtrate of fruit extract followed by addition of concentrated H_2SO_4. A yellow colouration observed in the extract indicated the presence of flavonoids.

2.3.4. Saponins (Frothing Test)

To 0.5 g of extract was added 5 ml of distilled water in a test tube. The solution was shaken vigorously and observed for a stable persistent froth. The frothing was mixed with 3 drops of olive oil and shaken vigorously. An appearance of creamy mass of small bubbles indicated the presence of saponins

2.3.5. Steroids

Two millimeter of acetic anhydride was added to 0.5 g of ethanol extract of each sample with 2 ml H_2SO_4. The colour changed from violet to blue or green in some samples indicating the presence of steroids.

2.3.6. Tannins

About 0.5 g of the extract was boiled in 10 ml of water in a test tube and then filtered. A few drops of 0.1 per cent ferric chloride was added and observed for brownish green or a blue-black colouration indicating the presence of tannins.

2.3.7. Terpenoids (Salkowski Method)

To 0.5 g each of the extract was added 2 ml of chloroform. Concentrated H_2SO_4 (3 ml) was carefully added to form a layer. A reddish brown colouration of the interface indicated the presence of terpenoids.

2.3.8. Cardiac Glycosides (Keller-Killiani test)

To 0.5 g of extract diluted to 5 ml in water was added 2 ml of glacial acetic acid containing one drop of ferric chloride solution. This was underplayed with 1 ml of concentrated sulphuric acid. A brown ring at the interface indicated the presence of deoxysugar characteristic of cardenolides. A violet ring may appear below the brown ring, while in the acetic acid layer a greenish ring may form just above the brown ring and gradually spread throughout this layer.

2.3.9. Isolation and Confirmation of Alkaloid

The dried powered was extracted with 10 per cent acetic acid in ethanol, the sample was left for atleast four hours. The extract was concentrated to one quarter of the original volume of the solution was precipitated with dropwise addition of ammonia solution followed by centrifugation and washing it with 1 per cent ammonia solution, the residue was dissolved in chloroform (Harbone, 1973).

Conformation of Alkaloids

The presence of alkaloids was confirmed using three reagents Dragnedroff's reagent, Mayer's reagent, Wagner's reagent, UV-Visible spectrophotometer and by TLC analysis.

UV-Visible spectrophotometer Analysis. The isolated alkaloids were confirmed by UV-Visible spectrophotometer using chloroform as a control.

2.4. Antioxidant Assays

2.4.1. Scavenging Activity of DPPH Radical

Scavenging activity of isolated alkaloid against DPPH radicals was assessed according to the method of Larrauri *et al.*, 1998 with some modifications. Briefly, 0.1 mM DPPH-methanol solution was mixed with 1 ml of 0.1mM DPPH methanol solution. After the solution was incubated for 30 min at 25° C in dark, the decrease in the absorbance at 517 nm was measured. Control contained methanol instead of antioxidant solution while blanks contained methanol instead of DPPH solution in the experiment. Ascorbic acid and BHT were used as positive controls.

The inhibition of DPPH radicals by the samples was calculated according to the following equation:

DPPH-scavenging activity (per cent) = [1-(absorbance of the sample-absorbance of blank)/absorbance of the control] ×100

2.4.2. Metal Chelating Activity

The chelation of ferrous ions by the extract was estimated by the previous method (Dinis *et al.*, 1994) with slight modification and compared with BHT and ascorbic acid. The chelation test initially includes the addition of ferrous chloride. The antioxidants present in the samples chelates the ferrous ions from the ferrous chloride. The remaining ferrous combine with ferrozine to form ferrous-ferrozine complex. The intensity of the ferrous-ferrozin complex formation depends on the chelating capacity of the sample and the colour formation was measured at 562 nm (Shimadzu UV-V is 2450).

0.1 µl of isolated alkaloid were added to a solution of 100 µl $FeCl_2$ (1 mM). The reaction was initiated by the addition of 250 µl ferrozine (1 mM). The mixture was finally quantified to 1.3 ml with methanol, shaken vigorously and left standing at room temperature for 10 min. after the mixture had reached equilibrium, the absorbance of the solution was measured spectrophotometrically. All the test and analysis were done in duplicate and average values were taken. The percentage inhibition of ferrous-ferrozine complex formation was calculated using the formula;

Per cent = 1-As/Ac X 100

where,

'Ac' is the absorbance of the control, 'As' is the absorbance of the sample.

3. Result and Discussion

The phytochemical investigation of methanol extract of *Hybanthus enneaspermus* leaves showed the presence of alkaloid, glycosides, flavonoids, tannins and steroid (Amutha priya *et al.*, 2011). In our present study, the qualitative phytochemical analysis was done by using colour forming and precipitating chemical reagents to generate preliminary data on the constituents of the plant extract. The chemical tests revealed the presence of phytochemical in the methanolic leaf extracts of *Hybanthus enneaspermus* (Table 12.1).

Table 12.1: Result of preliminary phytochemical analysis.

Sl.No.	Phytoconstituents	Observation	Methanol Extract
1.	Alkaloid	Colour change was observed in three reagents	+
2.	Flavonoids	Yellow colour formed	+
3.	Saponin	No froth form	−
4.	Steroids	Red colour change was observed	+
5.	Tannins	Brown-green colour was formed	+
6.	Terpenoids	No colour change	−
7.	Cardiac glycosides	Reddish brown colour ring was formed	+

3.1. Isolation of Alkaloid

The isolated alkaloid was stored and used for further analysis.

Confirmation of Alkaloid

The isolated alkaloid showed orange, white and yellow precipitate when tested with Dragnedroff, Mayer's and Wagner's reagent.

UV-Visible Spectrophotometer

The alkaloid when detected in UV-Visible spectrophotometer, the compound falls in the range of 263 nm (Figure 12.1). Molnár *et al.*, 1995 reported the alkaloid Navelbine at 258 nm which conform the presence of alkaloid.

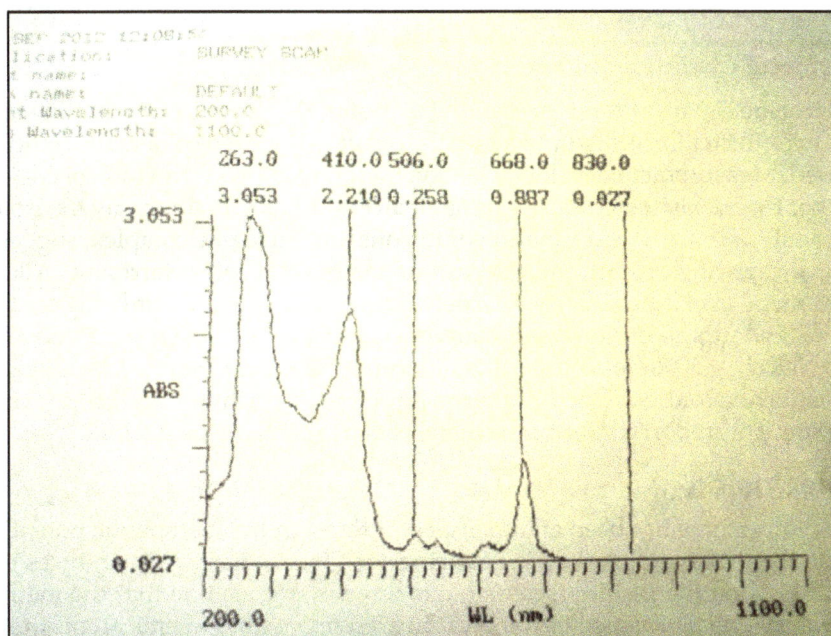

Figure 12.1: UV-Visible spectrophotometer analysis of crude alkaloid from *Hybanthus enneaspermus*.

3.2. Antioxidant Assays

3.2.1. Scavenging Activity of DPPH Radical

Free radical scavenging is one of the known mechanism by which antioxidants inhibit lipid peroxidation (Bloknina *et al.*, 2003; Evanas *et al.*, 1997). The DPPH radical scavenging activity has been extensively used for screening antioxidants from the extracts. When DPPH reacts with an antioxidant compound, which can donate hydrogen, it is reduced. The changes in colour (from deep-violet to light-yellow) where measured at 517 nm wavelength. As antioxidants donate protons to these radicals, the absorption decreases. The decrease in absorption is taken as a measure of the extent of radical scavenging. The result was observed as (46.97 per cent) that the isolated alkaloid of *Hybanthus enneaspermus* have a strong hydrogen-donating capacity and can efficiently scavenge DPPH radicals (Table 12.2). Amutha priya *et al.*, 2011 has reported 76 per cent of antioxidant activity in crude extract of *Hybanthus enneaspermus*. From our result the isolated alkaloid exhibited 46.97 per cent which shows alkaloid have a strong scavenging activity against DPPH radicals.

Table 12.2: Antioxidant activity of isolated alkaloid from *Hybanthus enneaspermus*.

Sl.No.	Antioxidant Assays	Standards		Sample (Alkaloid isolated from Hybanthus enneaspermus
		BHT	Ascorbic Acid	100µl
1.	DPPH assay	95.8	96.6	46.97 per cent
2.	Metal chelating	37.1	15.2	47.7 per cent

3.2.2. Metal Chelating Activity

Ferrozine can form complexes with Fe^{2+} but in the presence of chelating agent, the complex formation is disrupted with the result that the red color at the complex is decreased. Measurement of color reduction therefore allows estimation of chelating activity of the coexisting chelator (Yamaguchi *et al.*, 2000). In this assay the isolated alkaloid interfere with the formation of ferrous and ferrozine complex, suggesting they have chelating activity and can capture ferrous ion before ferrozine. Alkaloid showed metal chelating activity 47.7 per cent which are comparably higher to the positive standards ascorbic acid and butylated hydroxyl toluene respectively (Table 12.2). Kolak *et al.*, 2006 isolated six alkaloid from *Delphinium linearilobum* among that lycoctonine alkaloid exhibited higher metal chelating activity of 92.6 per cent at 100 µg/ml. This report conforms the metal chelating activity of alkaloid.

4. Conclusion

Recently, there has been an upsurge of interest in the therapeutic potential of medicinal plants as antioxidant in reducing such free radical which induces tissue injury. Based on the result discussed above we may conclude that the methanol extract of *Hybanthus enneaspermus* shows the presence of phytochemical constitutents and the alkaloid isolated possess significantly free radical scavenging activity. Further studies are required to idientify the alkaloid and to study the other antioxidant and anticancer activity of the leaf extract of *Hybanthus enneaspermus*.

References

Attda, Molnár, Káoly, Liliom, Ferenc, Orosz, Beáta, G., Vértessy and Judit Ovád, 1995. Anti-calmodulin potency of indol alkaloids in *in vitro* systems. *European Journal of Pharmacology*, 291: 73–82.

Amutha Priya, Ranganayaki, S. and Suganya Devi, P., 2011. Phytochemical screening and antioxidant potential of *Hybanthus enneaspermus*: A rare ethano-botanical herb. *Journal of Pharmacy Research*, 4(5): 1497–1502.

Blokhina, O., Virolainen, E. and Fagerstedt, K.V., 2003. Antioxidant, oxidative damage and oxygen deprivation stress: A review. *Annals of Botany*, 91: 179–194.

Chanda, S. and Dave, R., 2009. *In vitro* models for antioxidant activity evaluation and some medicinal plants possessing antioxidant properties: An overview. *African Journal of Microbiology*, 3(13): 981–996.

Dinis, T.C.P., Madeira, V.M.C. and Almeida, L.M., 1994. Action of phenolic derivatives (acetoaminophen, salicylate, and 5-aminosalicylate) as inhibitors of membrane lipid peroxidation and as peroxyl radical scavengers. *Arch. Biochem. Biophys*, 315: 161–169.

Halliwell, B. and Gutteridge, J.M.C., 2007. *Free Radical in Biochemistry and Medicine*, 4th edn. Oxford University Press, Oxford.

Harborne, J.B., 1973. *Phytochemical Methods: A Guide to Modern Technique of Plant Analysis*. Chapman and Hall Ltd., London, pp. 49–188.

Hemalatha, S., Wahi, A.K., Singh, P.N. and Chansouria, J.P.N., 2003. Anticonvulsant and free radical scavenging activity of *Hybanthus enneaspermus*: A preliminary screening. *Indian Journal Traditional Knowledge*, 2(4): 389.

Hill, A.F., 1952. *Economic Botany: A Textbook of Useful Plants and Plant Products*. McGraw-Hill Book Company Inc., NY.

Ibrahim, M., Hameed, A.J. and Jalbout, A., 2008. Molecular spectroscopic study of River Nile sediment in the Greater Cairo Region. *Applied Spectroscopy*, 62(3): 306–311.

Kirtikar, K.R. and Basu, B.D., 1991. In: *Indian Medicinal Plants*, Vol. 1, 2nd edn. Periodical Experts Book Agency, Delhi, pp. 212–213.

Larrauri, J.A., Sanchez-Moreno, C. and Saura-Calixo, F., 1998. Effect of temperature on the free radical scavenging capacity of extracts from red and white grape pomace peels. *Journal of Agricultural and Food Chemistry*, 46: 2694–2697.

Mandal, P., Misra, T.K. and Singh, I.D. 2010. Antioxidant activity in the extracts of two edible aroids. *Indian J. Pharm. Sci*, 72(1): 105–108.

Maryam, Zahin, Farrukh, Aqil and Iqbal, Ahmad, 2009. *In vitro* antioxidant activity and total phenolic content of four Indian medicinal plants. *Int. J. Pharm. Sci.*, 1(Suppl 1): 89–95.

Patel, D.K., Kumar, R., Prasad, S.K., Sairam, K. and Hemalatha, S., 2011. Antidiabetic and *in vitro* antioxidant potential of *Hybanthus enneaspermus* (Linn) F. Muell in streptozotocin-induced diabetic rats. *Asian Pacific Journal of Tropical Biomedicine*, 1: 316–322.

Rajakaruna, N., Harris, C.S. and Towers, G.H.N., 2002. Antimicrobial activity of plants collected from Serpentine outcrops in Sri Lanka. *Pharm. Bio.*, 40: 235–244.

Rice-Evans, C., Miller, N.J. and Paganga, G., 1997. Antioxidant properties of phenolic compounds. *Trends in Plant Science*, 2: 152–159.

Sasikumar, J.M., Gincy Marina, Mathew and Teepica Priya, Darsini, 2010. Comparative studies on antioxidant activity of methanol extract and flavonoid fraction of nyctanthes arbortristis leaves. *EJEAF Che*, 9(1): 227–233.

Sofowora, L.A., 1993. *Medicinal Plants and Traditional Medicine in Africa*. Spectrum Books Ltd., Ibadan, pp. 55–71.

Subhashini, R., Mahadeva Rao, U.S., Sumathi, P. and Gayathri, Gunalan, 2010. A comparative phytochemical analysis of cocoa and green tea. *Indian Journal of Science and Technology*, 3(2): 188–192.

Trease, G.E. and Evans, W.C., 1989. *Pharmacognosy*, 13th edn. Bailliere Tindall, London.

Weniger, B., Lagnika, L., Vonthron-Sénécheau, C., Adjobimey, T., Gbenou, J. and Moudachirou, M., 2004. Evaluation of ethnobotanically selected Benin medicinal plants for their *in vitro* antiplasmodial activity. *J. Ethnopharmacol.*, 90(2): 279–284.

Ufuk, Kolak, Mehmet, Ozturk, Fevzi, Ozgokce and Ayhan, Ulubelen, 2006. Norditerpene alkaloids from Delphinium linearilobum and antioxidant activity. *Phytochemistry*, 67: 2170–2175.

Yoganarasimhan, S.N., 2000. In: *Medicinal Plants of India–Tamil Nadu*, Vol. II. (Bangalore: Cyber Media), p. 276.

Scientific Basis of Herbal Medicine (2013)
Editor: Dr. Parimelazhagan Thangaraj
Published by: DAYA PUBLISHING HOUSE, NEW DELHI

Pages 115–126

Chapter 13

Volatile Oil Composition, Phytochemical Analysis and Antimicrobial Activity of *Syzygium caryophyllatum* (L.) Alston Leaves

N. Stalin and P.S. Swamy

Department of Plant Sciences, School of Biological Sciences,
Madurai Kamaraj University, Madurai – 625 021, Tamil Nadu, India

1. Introduction

The wild edible plants are rich source of phytochemicals, such as flavonoids, carotenoids and other phenolic compounds having high free-radical scavenging activity, helps to reduce the risk of chronic diseases, such as cardiovascular disease, cancer, and age related neuronal degeneration (Ames *et al.*, 1993). Colourful fruits are possibly a rich source of many dietary phenolic antioxidants and are believed to play an important role in the prevention of many oxidative and inflammatory diseases (Arts and Hollman, 2005). Natural products of higher plants may possess a new source of antimicrobial agents with possibly novel mechanism of action (Ahmad *et al.*, 2007 and Barbour *et al.*, 2004).They are effective in the treatment of infectious diseases while simultaneously mitigating many of the side effects that are often associated with synthetic antimicrobials (Iwu *et al.*, 1999).

Syzygium caryophyllatum (L.) Alston., commonly known as Wild black plum, a tropical evergreen tree belongs to the family Myrtaceae. It is a medium size tree distributed in Peninsular India mainly in the Western Ghats. Based on the threat perception due to its habitat loss and biotic pressure this tree is listed under the endangered category of IUCN Red List. The whole plant is used to cure variety of ailments in traditional medicine. Stem bark possess high medicinal value for treating diabetes mellitus (Ediriweera and Ratnasooriya, 2009) and shows antioxidant property. Essential oil of *Syzygium* species are reported to possess antioxidant, antibacterial (Chaieb *et al.*, 2007) activities. As plant is used for medicinal purposes, the investigation of chemical constituents in the plant parts and its biological activities may provide valuable information. No systematic studies on phytochemistry, antioxidant and antimicrobial activities are available in the literature. Therefore the present study focuses on the GC/MS analysis of the leaf volatile oil, phytochemical investigation, antioxidant activity and antimicrobial activity of *S. caryophyllatum*.

2. Materials and Methods

2.1. Plant Material and Extraction of Oil

The fresh leaves of *S.caryophyllatum* were harvested from Thirunelveli district of Tamilnadu and washed thoroughly with water. Hydrodistillation was carried out using a Clevenger type apparatus for 4 hours to obtain volatile oil from leaves. The oil was stored in screw-caped bottle and refrigerated until further analysis. For methanol and aqueous extractions 10 g of air-dried powder was taken in 100 ml of methanol and water in a conical flask plugged with cotton wool and then kept on a rotary shaker for 24 h. After 24 hours the extracts was filtered through Whatman No.1 filter paper and the filtrate was evaporated under reduced pressure. Dried extracts were stored in 4°C until use.

2.2. Analysis of Volatile Compounds

The chemical composition of the hydrodistilled volatile oil from the fresh leaves of *S. caryophyllatum* (L.) Alston was analyzed by GC/MS. A Shimadzu QP-2010 plus with thermal desorption system TD 20 was used to obtain the chromatograms. The name and specification of the column used is Omega wax™ 250 (30 m X 25 mm X 25 μm film thickness). The temperature was programmed from 80°C with 2 minute initial hold to 200°C at 4°C/minute and 200°C -235°C at 8° C/minute and a final hold for 9 minutes at 235°C. The injector and detector temperature were set at 270°C and 250°C at respectively and the split ratio was 1/60. Helium was used as the carrier gas and the ionizing voltage used is 70 eV. The components were identified based on the library search carried out using NIST and WILEY libraries.

2.3. Phytochemical Analysis

2.3.1. Estimation of Total Phenol Content

The amount of total phenol was determined with the Folin–Ciocalteu reagent using the method given by Lister and Wilson, (2001). This method was employed to evaluate the phenol content of the samples. A standard curve was prepared by using gallic acid as a standard. Different concentrations of gallic acid were prepared in 80 per cent of methanol, and their absorbances were recorded at 760 nm. 100 μl of sample was dissolved in 500 μl (1/10 dilution) of the Folin-Ciocalteu reagent and 1000 μl of distilled water. The solutions were mixed and incubated at room temperature for 1 min. After 1 min, 1500 μl of 20 per cent sodium carbonate (Na_2CO_3) solution was added. The mixture was shaken and then incubated for 2 h in the dark at room temperature. The absorbances of all samples were measured at 760 nm using a UV–V is spectrophotometer (Model. U.2800, Hitachi) and the results are expressed in mg of gallic acid equivalents (GAE) per g of dry weight of the plant. The amount of phenol in plant extracts in gallic acid equivalents (GAE) was calculated by the following formula:

X = (A. mo)/(Ao.m)

where,

X is the phenol content, mg/mg plant extract in GAE, A is the absorption of plant extract solution, Ao is the absorption of standard gallic acid solution, m is the weight of plant extract, mg and mo is the weight of gallic acid in the solution, mg.

2.3.2. Estimation of Total Flavonoid Content

The flavonoid content in extracts was determined spectrophotometrically followed by Quettier-Deleu *et al.* (2000) using a method based on the formation of a complex flavonoid–aluminium, having the absorbtivity maximum at 430 nm. Rutin was used to make the calibration curve. 1 ml of diluted sample was separately mixed with 1 ml of 2 per cent aluminum chloride methanolic solution. After incubation at room temperature for 15 min, the absorbance of the reaction mixture was measured at 430 nm with a UV–Vis spectrophotometer (Model. U. 2800, Hitachi) and the flavonoid content was expressed in mg per g of rutin equivalent (RE). The amount of flavonoid in plant extracts in RE was calculated by the same formula as that of phenol:

X = (A. mo)/(Ao.m)

2.3.4. Estimation of Tannin Content

The amount of tannin content was determined by modified Prussian blue method (Shanmugam *et al.*, 2010). This method is based on the mechanism that the phenols reduce potassium ferricyanide to produce ferrous ions; these ferrous ions in turn react with ferric chloride in the presence of dilute HCl to form a Prussian blue coloured complex, which can be measured spectrophotometrically at 700 nm wavelength. Ascorbic acid was used to make a calibration curve. The amount of tannin in plant extracts in tannic acid equivalent (TAE) was calculated by the same formula as that of phenol.

2.4. *In vitro* Antioxidant Assay

2.4.1. Free-Radical Scavenging Ability (DPPH-assay)

The scavenging ability of methanol extract on 1,1-diphenyl-2-picrylhydrazyl (DPPH) free radicals was estimated by Shimada *et al.* (1992). This method depends on the reduction of purple DPPH to a yellow coloured diphenyl picrylhydrazine and the remaining DPPH which showed maximum absorption at 517 nm was measured. About 2 ml of various concentrations of test samples were mixed with 0.5 mL of 1 mM DPPH in methanol. An equal amount of methanol and DPPH served as control. The mixture was shaken vigorously and then steadily stayed for 30 min at room temperature in dark. The absorbance of the resulting solution was measured at 517 nm against the blank using a UV–Vis spectrophotometer (Model. U. 2800, Hitachi). The experiment was performed in triplicates. The higher the scavenging ability value, the higher the antioxidant activity is detected in the methanol extract (Shimada *et al.*, 1992). The DPPH radical scavenging activity was calculated according to the following equation:

Per cent DPPH radical scavenging activity = (Ao – A1)/Ao × 100 per cent *where,*

Ao is the absorbance of the control reaction and A1 is the absorbance in the presence of the sample of the tested extracts. Percentage of radical activity was plotted against the corresponding antioxidant substance concentration to obtain the IC_{50} value, which is defined as the amount of antioxidant substance required to scavenge the 50 per cent of free radicals present in the assay system. IC_{50} values are inversely proportional to the antioxidant potential.

2.4.2. Determination of Total Antioxidant Capacity by Phosphomolybdenum Method

The total antioxidant capacities of the fruit extract of *S. caryophyllatum* was evaluated by the phosphomolybdenum method as described by Prieto, Pineda, and Aguilar (1999). The assay is based on the reduction of Mo(VI) to Mo(V) by the extract and subsequent formation of a green phosphate/Mo(V) complex at acidic pH. 0.3 ml of each sample solution and ascorbic acid (100 µg/ml) were combined with 3 ml of reagent (0.6 M sulphuric acid, 28 mM sodium phosphate and 4 mM ammonium molybdate). A typical blank solution contained 3 ml of reagent solution and the appropriate volume of the same solvent used for the sample. All tubes were capped and incubated in a boiling water bath at 95°C for 90 min. After the samples had been cooled to room temperature, the absorbance of the solution of each sample was measured at 695 nm against the blank using a UV–Vis spectrophotometer (Model. U.2800, Hitachi). The experiment was performed in triplicates. The antioxidant activity is expressed as the number of equivalents of ascorbic acid.

2.5. Antimicrobial Sensitivity Tests

Antibacterial and antifungal activities of methanol and aqueous plant extracts were investigated by the agar well diffusion method described by Irobi *et al.* (1994). Mueller-Hinton agar medium (MHA) for bacteria and potato dextrose agar medium (PDA) for fungus respectively were used for antibacterial and antifungal susceptibility tests. For agar well diffusion method, a well was prepared in the plates with the help of a cork-borer (0.5 cm). 25 µL of each extract at a concentration of 5.0 and 10.0 mg/mL for fungi and bacteria were introduced into the well. DMSO serve as a control, and antibiotic disks (5.0 mm in diameter) of 10/10mcg/ml Ampicillin/Sulbactum, 120mcg/ml Gentamycin (for bacteria), and Nystatin (for fungi) were also used as positive controls. All the plates were incubated at 37°C for 18-20 hrs for bacteria and at 28°C for 48-96 hrs for fungi. The zones of growth inhibition were measured after 18-20 hrs of incubation at 37°C for bacteria and 48-96 hrs for fungi at 28°C, respectively. The sensitivity of the microorganism species to the plant extracts was determined by measuring the sizes of the inhibitory zones. All the experiments were performed in triplicate and the mean values are presented.

2.6. Statistical Analysis

All the experiments were repeated three times and the results were reported as mean±SD (standard deviation of estimate). The IC_{50} values were calculated using the ED_{50} plus v 1.0 programme. Statistical analysis was performed using Microsoft Excel.

3. Results and Discussion

3.1. Chemical Composition of the Essential Oil

The volatile oil was obtained by conventional hydridistillation of *S. caryophyllatum* leaves in a clevenger-type apparatus, which gave oil in pale yellow colour (0.2 per cent v/w of fresh weight). Fifty five compounds representing 99.98 per cent of the leaves oil were identified (Table 13.1). α-Cadinol (24.68), Myristicin (12.02 per cent) δ-Cadinene (8.4 per cent) were the most prominent compounds followed by β- Caryophyllene (5.33 per cent), α-Cadinol (5.11 per cent), β-cis-ocimene (2.53 per cent), β-Elemen (2.52 per cent) and Diepi alpha cedrene epoxide (2.31 per cent). as the major compound. The volatile oil of *S. caryophyllatum* contains antimicrobial components such as β-caryophyllene, caryophyllene oxide a common sesquiterpene possesses anti-inflammatory and anticarcinogenic activities (Zheng, 1992) and α-Cadinol possess anti fugal activity (Hui-Ting Chang, 2008). Myristicin is used as a fragrance in the cosmetic industry and as a flavouring agent in food (Slamenova *et al.*, 2009). In traditional medicine, myristicin has been used to treat cholera, stomach cramps, nausea, diarrhea, and anxiety (Martins, 2011). Recently, it was reported that myristicin has anti-inflammatory properties related with its inhibition of NO, cytokines, chemokines and growth factors dsRNA stimulated macrophages via the calcium pathway (Ji Young Lee and Wansu Park, 2011) and it also exerts antibacterial activity against Gram-positive and Gram-negative organisms (Narasimhan and Dhake, 2006).

Table 13.1: Chemical composition of volatile oil of *Syzygium caryophyllatum* (L.) Alston.

Sl.No.	R. Time	Area per cent	Name
1.	4.467	0.34	β-Trans-ocimene
2.	4.801	2.53	β-Cis-ocimene
3.	9.642	0.28	α-Cubebene
4.	10.596	0.64	Copaene
5.	11.859	1.9	β-Cubebene
6.	13.123	0.42	α-Bergamotene
7.	13.276	2.52	β-Elemen-(2)
8.	13.486	5.33	β-Caryophyllene
9.	14.596	0.64	Germacrene B
10	15.199	0.18	β-Cadinene
11.	15.484	1.38	α-Caryophyllene
12.	15.986	0.7	α-Amorphene
13.	16.186	0.52	δ-Guajene
14.	16.564	1.77	Germacrene D
15.	16.704	1.11	α-Bulnesene
16.	16.974	1.5	α-Muurolene
17.	17.227	1.02	Bicyclogermacrene

Contd...

Table 3.3–Contd...

Sl.No.	R. Time	Area per cent	Name
18.	17.852	8.4	δ-Cadinene
19.	18.298	1.01	α-Curcumene
20.	18.526	0.19	Cadina-1,4-diene
21.	18.789	0.27	α-Muurolen
22.	19.742	1.15	Germacrene B
23.	21.269	0.21	Cubebol
24.	23.684	0.3	Caryophyllene oxide
25.	24.041	0.23	Viridiflorol
26.	24.801	0.35	Ledol
27.	25.415	0.4	Cubeban-11-ol
28.	25.514	1.3	Cubenol
29.	25.668	1.69	Epicubenol
30.	25.904	1.83	Globulol
31.	26.109	1.37	Viridiflorol
32.	26.647	0.56	Rosifoliol
33.	26.898	1.1	Cedrol
34.	27.077	0.37	Spathulenol
35.	27.39	2.31	Diepi-alpha-cedrene epoxide
36.	27.649	0.25	Globulol
37.	28.178	5.11	τ-Cadinol
38.	28.54	6.65	α-Cadinol
39.	28.804	12.02	Myristicin
40.	29.03	1.81	Ledol
41.	29.323	0.51	α-Eudesmol
42.	29.561	18.3	α-Cadinol
43.	30.016	0.94	Junipercamphor
44.	30.2	0.29	Heptadecanoate
45.	31.007	0.5	Junipercamphor
46.	31.233	0.47	Cycloheptane, 4-methylene-1-methyl-2-(2-methyl-1-propen-1-yl)-1-vinyl-
47.	33.194	1.54	(-)-Isolongifolol, methyl ether
48.	35.597	0.45	Ethyl oleate
49.	37.085	1.21	Linoleic acid ethyl ester
50.	38.946	0.21	Linolenate <methyl->
51.	39.357	0.3	Phytol
52.	39.835	0.77	Methyl 18-methylnonadecanoate
53.	40.38	0.46	Heptadecyl trifluoroacetate
54.	43.797	1.4	Methyl docosanoate
55.	49.113	0.97	Methyl lignocerate

3.2. Total Phenol, Flavonoid and Tannin Contents

The quantitative estimation of total phenol, flavonoid and tannin content of the methanol and aqueous extracts of *S. caryophyllatum* leaves are shown in Figure 13.1. Total phenol content in terms of gallic acid equivalent (GAE) were between 530.71 mg/g and 156.05 mg/g dry material and total flavonoid content in terms of rutin equivalent (RE) were between 139.74 mg/g and 44.58 mg/g dry weight. While the tannin content of the methanol and aqueous extracts expressed as tannic acid equivalent were 733.28 mg/g and 348.57 mg/g dry weight respectively. The above results showed that methanol extract has more potent than aqueous extract. Tannins were reported to exhibit antiviral, antibacterial and anti-tumour activities. According to recent reports, a highly positive relationship between phenols and antioxidant activities showed in *Syzygium aqueum* species (Hasnah *et al.*, 2009).

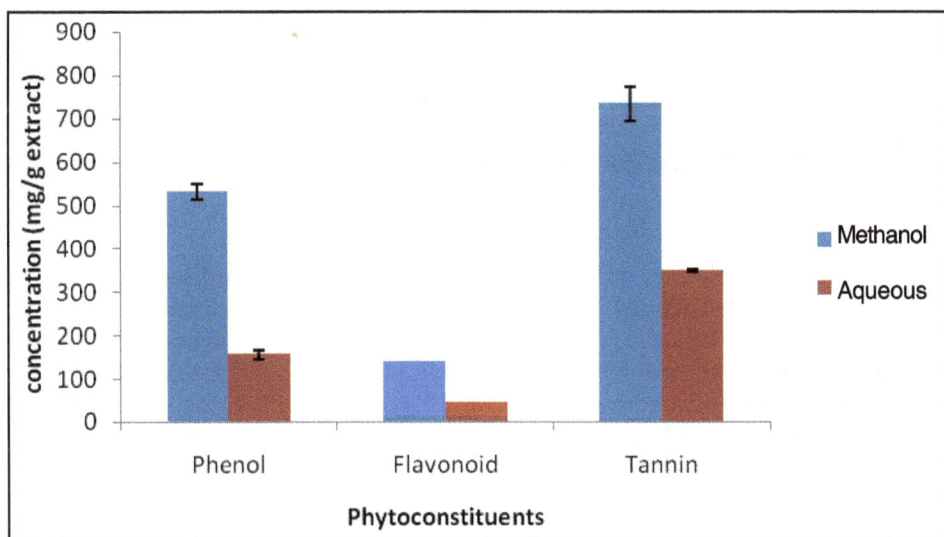

Figure 13.1: Quantitative estimation of total phenol, flavonoid and tannin content of *Syzygium caryophyllatum* (L) Alston.

3.3. Evaluation of Antioxidant Activity and Scavenging Ability

The fruit extract was used to determine the total antioxidant capacity and free radical scavenging activity. DPPH free radical activities of methanol fruit extract and standard compound were evaluated and the results are shown in Figure 13.2. Free radical scavenging activity indicated that methanol fruit extract has significant radical scavenging ability on DPPH with IC50 = 70.47±2.0 µg/ml, which is close to the positive control ascorbic acid (IC50 = 41.51±10.8 µg/ml). DPPH is a purple colour stable free radical and it transform to yellow colour non radical by abstracting one electron, hence it is widely used as to measure the electron donating capacity of the antioxidant under the assayed condition (Molyneux *et al.*, 2004). Free radicals are involved in many disorders like neurodegenerative diseases, cancer and AIDS. Antioxidants through their scavenging power are useful for the management of these

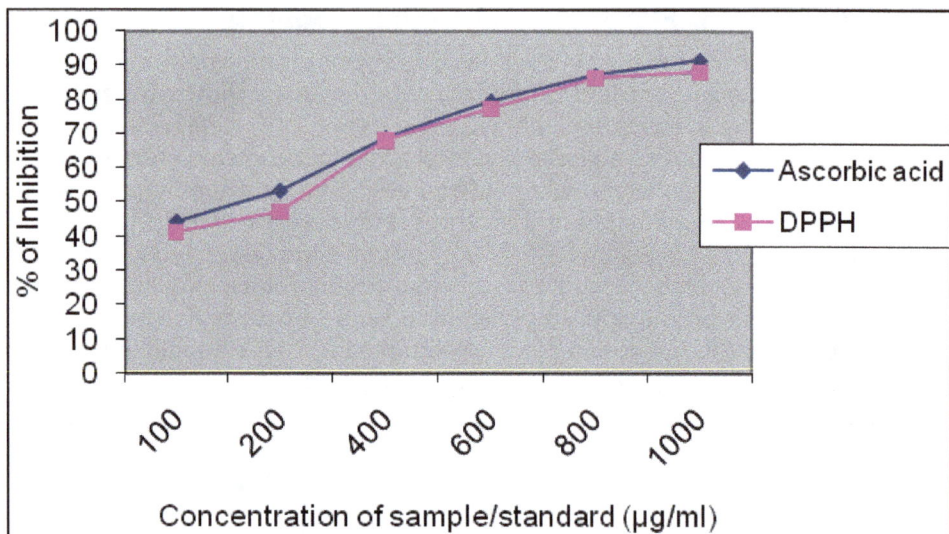

Figure 13.2: DPPH-free radical scavenging activities of methanol fruit extract and ascorbic acid.

diseases. DPPH stable free radical method is an easy, rapid and sensitive way to survey the antioxidant activity of a specific compound or plant extracts (Koleva *et al.*, 2002). The total antioxidant capacity was done by the phosphomolybdenum method. This method is based on the reduction of Mo (VI) to Mo (V) by the antioxidant compounds and the formation of a green Mo (V) complex with a maximal absorption at 695 nm (Kumaran and Karunakaran, 2006). The methanol fruit extract of *Syzygium caryophyllatum* showed significant antioxidant activity (454±28.4 mg/g of ascorbic acid equivalent). Many natural antioxidants are plant phenolics found in every part of the plant, including the leaves, fruits and seeds (Kim *et al.*, 1997). According to Zhang and Lin (2009) antioxidant activity increased proportionally with tannins content.

3.4. Antimicrobial Sensitivity Test

The antimicrobial activity of *S. caryophyllatum* methanol and aqueous leaves extracts was assayed *in vitro* by agar well diffusion method against 15 microorganisms which consisted of 7 bacteria and 8 fungi species. The results of the methanol and aqueous extract of leaves exhibited antibacterial activity against all the tested strains *viz.* *B. cereus, B. licheniformis, E. coli, S. aureus, S. hominis, A. viridans* and *K. pneumoniae* as shown in Table 13.2. Both the leaf extracts exhibited the pronounced antibacterial activity against *S. aureus, E.coli, K. pneumoniae, A.viridans* and *B.cereus* than the other strains tested. The zones of inhibition were ranging from 7-18 mm in diameter. The highest zone of inhibitions (18 mm) noted in methanol extract against *S. aureus* in 10 mg/ml concentration. Methanol extracts were more active than the aqueous extract against all the microorganisms. The results of the antifungal activity shown in Table 13.3, the methanol and aqueous extract showed marked antifungal activity against

Table 13.2: Antibacterial sensitivity test using agar well diffusion method.

Organisms	Zone of Inhibition (mm) including Well Size 5mm					
	MeOH Extract		Aqueous Extract		Amp/Sulb (10/10mcg/ ml)	Gentamycin (120 mcg/ ml)
	5 mg/ml	10 mg/ml	5 mg/ml	10 mg/ml		
Staphylococcus aureus	12±1.0	18±1.0	11.7±0.5	13.7±1.5	13.5±1.5	21±1.1
S.hominis	9±1.0	11±1.1	–	7±0.5	30±0.5	28±0.5
K. pneumoniae	11±0.5	14.6±0.5	9.7±1.5	11.7±1.5	10±0.5	22±0.5
A. viridans	11.5±0.5	14±0.5	9.5±1.1	11±1.0	10±1.5	24.5±0.5
Escherichia coli	12.5±0.5	15±1.0	9±1.0	10±0.5	10±2.0	25±0.5
Bacillus cereus	11.5±0.5	14±0.5	9.5±1.1	11.5±0.5	9.5±1.1	27±1.1
B.licheniformis	7.5±0.5	8.0±0.5	–	7.5±0.5	31±1.1	23±1.0

Table 13.3: Antifungal and antibacterial sensitivity test using well diffusion method.

Organisms	Zone of Inhibition (mm) including Well Size 5mm				
	MeOH Extract		Aqueous Extract		Nystatin (50 mcg/ml)
	5 mg/ml	10 mg/ml	5 mg/ml	10 mg/ml	
Aspergillus flavus	15±1.5	17.5±1.5	11.5±0.5	13.5±0.5	19±1.1
A.niger	12.5±2.1	16±2.1	7±0.5	8.5±1.5	25±1.0
A.fumigatus	15±3.6	18±2.9	10±0.5	14±1.0	27.5±2.1
Penicillium murneffei	17±3.0	19±2.5	10±1.0	12.7±1.5	27±2.1
Microsporum sp	10±1.0	13.5±0.5	–	–	15±1.0
Rhizopus sp	–	–	–	–	10±1.5
Epidemoshyton sp	–	–	–	–	27±1.0
Candida sp	–	–	–	–	21±2.6

A. flavus, A. niger, A. fumigatus and *P. murneffei* but the other fungi such as *Candida* sp., *Epidermoshyton* sp., *Microsporum* sp. and *Rhizopus* sp. did not show any antifungal activities. The blind control (DMSO) did not inhibit any of the microorganisms tested. Among the two extracts used in the antimicrobial susceptibility test, the methanol extract showed higher antibacterial and antifungal activity than that of aqueous extract. In general, aqueous extracts exhibited very low antimicrobial activities compared to the methanol extracts. This may be due to the solvent to extract the different constituents having antimicrobial activity (Chandrasekaran, 2004). Earlier, Chandrasekaran and Venkatesalu (2004) reported that methanol and aqueous seed extract of *S. cumini* fruit was very effective against certain fungal species, namely *A. flavus, A. fumigatus, and A. niger.* The antimicrobial activity of the *S. cumini* leaves methanol and aqueous extract may be due to tannins and other phenolic constituents. *S. cumini* is known to be very rich in gallic and ellagic acid polyphenol derivatives (Chattopadhyay, 1998 and Mahmoud, 2001). The extracts of higher plants can be very good source of antibiotics against various bacterial pathogens (Fridous *et al.,* 1990).

4. Conclusion

The results of the present study showed that the extracts of *S. caryophyllatum* contain high amount of tannins and phenolic compounds, which may be attributed to the good antimicrobial and antioxidant potential as shown. A potent scavenger of free radicals may serve as a possible preventative intervention for the diseases (Gyamfi *et al.,* 1999). Both the methanol and aqueous leaf extracts showed remarkable anti bacterial and antifungal activities. This may be related to the high amount of tannins and phenolic compounds in this plant extract. Thus the *S. caryophyllatum* leaf extracts exhibited promising antimicrobial and antioxidant activity suggesting further research towards bio-prospecting.

Acknowledgment

First author thank Mr. Ajay Kumar of Advanced instrumentation research facility, JNU, New Delhi for extending help in GC/MS analysis and also would like to thank UGC for the award of meritorious fellowship. PSS thanks the UGC for the funding of major research project (MRP) and also for partial financial support through UGC-CAS, DST-IPLS and DST-Purse programme.

References

Ahmad, I. and Aqil, F., 2007. *In vitro* efficacy of bioactive extracts of 15 medicinal plants against ESbL-producing multi-drug resistant enteric bacteria. *Microbiol. Res.,* 162: 264–275.

Ames, B.M., Shigena, M.K. and Hagen, T.M., 1993. Oxidants, antioxidant and the degenerative diseases of aging. *Proc. Natl. Acad. Sci., USA,* 90: 7915–7922.

Arts, I.C.W. and Hollman, P.C.H., 2005. Polyphenols and disease risk in epidemiologic studies. *Am. J. Clin. Nutr.,* 81(1): 317–325.

Bakkali, F., Averbeck, S., Averbeck, D. and Idaomar, M., 2008. Biological effects of essential oils: A review. *Food Chem. Toxicol.,* 46: 446–475.

Barbour, E.K., Al Sharif, M., Sagherian, V.K., Habre, A.N., Talhouk, R.S. and Talhouk, S.N., 2004. Screening of selected indigenous plants of Lebanon for antimicrobial activity. *J. Ethnopharmacol.,* 93: 1–7.

Chaieb, K., Hajlaoui, H., Zmantar, T., Kahla-Nakbi, A.B., Rouabhia, M., Mahdouani, K. and Bakhrouf, A., 2007. The chemical composition and biological activity of clove essential oil, *Eugenia caryophyllata* (*Syzigium aromaticum* L. Myrtaceae): A short review. *Phytother. Res.,* 21: 501–506.

Chandrasekaran, M. and Venkatesalu, V., 2004. Antibacterial and antifungal activity of *Syzygium jambolanum* seeds. *J. Ethnopharmacol.,* 91: 105–108.

Chang, H.T., Cheng, Y.H., Wu, C.L., Chang, S.T., Chang, T.T. and Su, Y.S., 2008. Antifungal activity of essential oil and its constituents from *Calocedrus macrolepis* var. *formosana* Florin leaf against plant pathogenic fungi. *Bioresource Technol.,* 99: 6266–6270.

Chattopadhyay, D., Sinha, B.K. and Vaid, L.K., 1998. Antibacterial activity of *Syzygium* species. *Fitoterapia*, 69: 356–367.

Dudonne, S., Vitrac, X., Coutiere, P., Woillez, M. and Merillon, J.M., 2009. Comparative study of antioxidant properties and total phenolic content of 30 plant extracts of industrial interest using DPPH, ABTS, FRAP, SOD, and ORAC assays. *J. Agric. Food Chem.*, 57: 1768–1774.

Ediriweera, E.R.H.S.S. and Ratnasooriya, W.D., 2009. A review on herbs used in treatment of *Diabetes mellitus* by Sri Lankan and traditional physicians. *Ayu.* 30(4): 373–391.

Fridous, A.J., Islam., S.N.L.M. and Faruque, A.B.M., 1990. Antimicrobial activity of the leaves of *Adhatoda vasica, Calatropis gigantium, Nerium odoratum* and *Ocimum sanctum. Bangladesh J. Bot.*, p. 227.

Gyamfi, M.A., Yonamine, M. and Aniya, Y., 1999. Free-radical scavenging action of medicinal herbs from Ghana *Thonningia sanguinea* on experimentally-induced liver injuries. *General Pharmacol.*, 32: 661–667.

Hallström, H. and Thuvander, A., 1997. Toxicological evaluation of myristicin. *Nat. Toxins.*, 5: 186–192.

Irobi, O.N., Moo-Young, M., Anderson, W.A. and Daramola, S.O., 1994. Antimicrobial activity of the bark of *Bridelia ferruginea* (Euphorbiaceae). *Int. J. Pharmacog.*, 34: 87–90.

Iwu, M.W., Duncan, A.R. and Okunji, C.O., 1999. New antimicrobials of plant origin. In: *Perspectives on New Crops and its Uses*, (Ed.) J. Janick. ASHS Press, Alexandria, V.A. pp. 457–462.

Jun, N.J., Mosaddik, A., Moon, J.Y., Jang, K.C., Lee, D.S., Ahn, K.S. and Cho, S.K., 2011. Cytotoxic activity of betacaryophyllene oxide isolated from jeju guava (*Psidium cattleianum* Sabine) leaf. *Rec. Nat. Prod.*, 5(3): 242–246.

Kahkonen, M.P., Hopia, A.I., Vuorela, H.J., Rauha, J.P., Pihlaja, K. and Kujala, T.S., 1999. Antioxidant activity of plant extracts containing phenolic compounds. *J. Agric. Food Chem.*, 47: 3954–3962.

Kim, B., Kim, J., Kim, H. and Heo, M., 1997. Biological screening of 100 plants for cosmetic use. II. Antioxidative activity and free radical scavenging activity. *Int. J. Cosmet. Sci.*, 19: 299–307.

Koleva, I.I., Van Beek, T.A., Linssen, J.P.H., de Groot, A. and Evstatieva L.N., 2002. Screening of plant extracts for antioxidant activity: A comparative study on three testing methods. *Phytochem. Anal.*, 13: 8–17.

Kumaran, A. and Karunakaran, J.R., 2006. *In vitro* antioxidant activities of methanol extracts of five *Phyllanthus* species from India. *LWT–Food. Sci. Technol.*, 40: 344–352.

Lee, B.K., Kim, J.H., Jung, J.W., Choi, J.W., Han, E.S., Lee, S.H., Ko, K.H. and Ryu, J.H., 2005. Myristicin–induced neurotoxicity in human neuroblastoma SK–N–SH cells. *Toxicol. Lett.*, 157: 49–56.

Lee, J.Y and Park, W., 2011. Anti-inflammatory effect of myristicin on RAW 264.7 macrophages stimulated with polyinosinic–polycytidylic acid. *Molecules*, 16: 7132–7142.

Mahmoud, I.I., Marzouk, M.S.A., Moharram, F.A., El-Gindi, M.R. and Hassan, A.M.K., 2001. Acylated flavonol glycosides from *Eugenia jambolana* leaves. *Phytochemistry*, 58: 1239–1244.

Martins, C., Doran, C., Laires, A., Rueff, J. and Rodrigues, A.S., 2011. Genotoxic and apoptotic activities of the food flavourings myristicin and eugenol in AA8 and X., RCC1 deficient EM9 cells. *Food Chem. Toxicol.*, 49: 385–392.

Molyneux, P., 2004. The use of the stable free radical diphenyl picryl hydroxyl (DPPH) for estimating antioxidant activity. *LWT Food Sci. Technol.*, 26: 211–219.

Narasimhan, B. and Dhake, A.S., 2006. Antibacterial principles from *Myristica fragrans* seeds. *J. Med. Food.*, 9: 395–399.

Osman, H., Rahim, A.A., Isa, N.M. and Bakhir, N.M., 2009. Antioxidant activity and phenolic content of *Paederia foetida* and *Syzygium aqueum. Molecules*, 14: 970–978.

Prieto, P., Pineda, M. and Aguilar, M., 1999. Spectrophotometric quantitation of antioxidant capacity through the formation of a phosphomolybdenum complex, Specific application to the determination of vitamin E. *Anal. Biochem.*, 269: 337–341.

Quettier, D.C., Gressier, B., Vasseur, J., Dine T., Brunet, C., Luyckx, M.C., Cayin, J.C., Bailleul, F. and Trotin, F., 2000. Phenolic compounds and antioxidant activities of buckwheat (*Fagopyrum esculentum* Moench) hulls and flour. *J. Ethnopharmacol.*, 72: 35–42.

Shanmugam, T., Sathish kumar, T. and Panneer Selvam, K., 2010. *Laboratory Handbook on Biochemistry*. PHI Learning Pvt. Ltd., New Delhi.

Shimada, K., Fujikawa, K. Yahara, K. and Nakamura, T., 1992. Antioxidative properties of xanthan on the autoxidation of soybean oil in cyclodextrin emulsion. *J. Agric. Food Chem.*, 40: 945–948.

Slamenova, D., Horvathova, E., Wsolova, L., Sramkova, M. and Navarova, J., 2009. Investigation of antioxidative, cytotoxic, DNA-damaging and DNA-protective effects of plant volatiles eugenol and borneol in human-derived HepG2, Caco-2 and VH10 cell lines. *Mutat. Res.*, 677: 46–52.

Yang, D., Michel, L., Chaumont, J. and Millet-Clerc, J., 1999. Use of caryophyllene oxide as an antifungal agent in an *in vitro* experimental model of onychomycosis. *Mycopathologia*, 148: 79–82.

Zhang, L.L. and Lin, Y.M., 2009. Antioxidant tannins from *Syzygium cumini* fruit. *Afr. J. Biotechnol.*, 8: 2301–2309.

Zheng, G., Kennedy, P.M. and Lam, L.K.T., 1992. Sesquiterpenes from clove (*Eugenia caryophyllata*) as potential anticarcinogenic agents. *J. Nat. Prod.*, 55: 999–1003.

Scientific Basis of Herbal Medicine (2013)
Editor: Dr. Parimelazhagan Thangaraj
Published by: DAYA PUBLISHING HOUSE, NEW DELHI

Pages 127–132

Chapter 14

Pharmacognostical Studies and Preliminary Phytochemical Investigations on the Rhizome of *Aponogeton natans* L. (Aponogetonaceae)

Somu Aron and Palanichamy Mehalingam

Research Department of Botany,
V.H.N. Senthikumara Nadar College (Autonomous),
Virudhunagar – 626 001, Tamil Nadu, India

1. Introduction

Aponogeton natans L (Family: Aponogetonaceae) is an aquatic tuberous herb. In traditional medicinal system, this plant is used to treat wound healing and also used to treat skin diseases (Yoganarasimhan, 2000). *Kani* tribe use it to treat cuts and wounds (Britto and Mahesh, 2007). The World Health Organization estimates that some 80 per cent of the people in developing world rely on traditional medicine and that to these 85 per cent use plants or their products as the remedies. India is endowed with an estimated 47,000 species of plant that include around 8000 plants which are known to have medicinal properties.

The pharmacognostical studies of this taxon have not been reported. Hence, the present work has been undertaken to establish the various pharmacognostical parameters, which could serve as a valuable source of information and provide suitable standards for the further identification of this plant.

2. Material and Methods

2.1. Collection of Specimens

The rhizome of *Aponogeton natans* was collected from Arapathy village, Ramanathapuram District, Tamilnadu. The authenticity of the plant was confirmed with The Flora of Presidency of Madras (Gamble, 1935). It was deposited at Department of Botany VHNSN College (Autonomous), Virudhunagar for future reference. Care was taken to select healthy plants and normal organs. The rhizomes were cut and fixed in FAA (Formalin-5ml + Acetic acid-5 ml + 70 per cent Ethyl alcohol-90 ml). After 24 hrs of fixing, the specimens were dehydrated with graded series of Tertiary-Butyl alcohol as per the schedule given (Sass, 1940). Infiltration of the specimens was carried by gradual addition of paraffin wax (melting point 58-60°C) until TBA solution attained super saturation. The specimens were cast into paraffin blocks.

2.2. Sectioning

The paraffin embedded specimens were sectioned with the help of Rotary Microtome. The thickness of the sections was 10-12 μm. Dewaxing of the sections was by customary procedure (Johansen, 1940). The sections were stained with Toluidine blue as per the method published (O'Brien *et al.*, 1964). Wherever necessary, the sections were also stained with Safranin and Fast green.

2.3. Photomicrographs

Microscopic descriptions of tissues were supplemented with micrographs wherever necessary. Photographs of different magnifications were taken with Nikon lab photo 2 microscopic Unit. For normal observations bright field was used. For the study of crystals, starch grains and lignified cells, polarized light was employed. Since these structures have birefringent property under polarized light, they appear bright against dark background. Magnifications of the figures are indicated by the scale-bars. Descriptive terms of the anatomical features were as given in the standard Anatomy books (Easu, 1964).

2.4. Physico-chemical Constants

Physico-chemical parameters of *Aponogeton natans* rhizome powder were determined and reported as total ash, water-soluble ash, acid-insoluble ash, alcohol-soluble extractive, water-soluble extractive and moisture content (Anonymous, 1985).

2.5. Preliminary Phytochemical Analysis

Shaded dried and powdered plant samples were successively extracted with ethanol, acetone and ethyl acetate. The extracts were filtered and concentrated using vacuum distillation. The different extracts were subjected to qualitative tests for the identification of various phytochemical constituents as per standard procedure (Harborne, 1998; Kokate *et al.*, 2003).

3. Results and Discussion

3.1. Macroscopic Characters

Tuber linear-oblong, sometimes globose, 2.5-3.5 × 0.8-1.5 cm. Leaves oblong, 3-11 × 0.8-2(3) cm, chartaceous, glabrous, base subcordate-obtuse, margin entire, apex obtuse subacute; petiole 6-40 cm. Spikes 2-7 × 0.3 cm; peduncle 10-45 cm, usually overtopping leaves. Flowers patent, dense throughout, ca. 4 mm across. Tepals 2, violetish, obovate-suborbicular 2.5 × 1.5 mm, nearly twice as long as broad, claw to 1mm. Stamens 6; filaments 1.5-3 mm; anthers to 0.5 mm. Ovaries 3, oblong, to 1mm; ovules ca. 8; style to 1.5 mm. Fruit globose, to 4 × 2 mm, smooth, sharply beaked; seeds ca. 8, terete, 1.8-2 × 0.3 mm; testa double: outer one loose-winged, transparent, reticulately nerved; inner one appressed to embryo; plumule obscure (Figure 14.1A).

3.2. Microscopic Character

3.2.1. Rhizome

The rhizome is circular in sectional view with distinct periderm and homogenous parenchymatous starch filled ground tissue (Figure 14.1B). The periderm is continuous and uniformly thick. The ground tissue is compact (Figure 14.1C).

3.2.2. Periderm

The epidermis is replaced by the periderm. It is four or five layered; the cells are thin walled, suberised and tabular in shape. Ground tissue is parenchymatous; the cells are wide, circular or angular and compact. The cell walls are thin. The cells have distinct nuclei; they are uninucleate or multinucleate (Figure 14.1D). Vascular strands are poorly developed. They have irregular mass of small vascular elements (Figure 14.1E). The xylem elements are few, narrow, thick walled and lignified (Figure 14.1F), Phloem elements are more in number than the xylem element and they occur scattered near the xylem. The vascular elements ramify in the rhizome to different regions (Figure 14.1G).

When the sections of the rhizome are viewed under the polarized light microscope, the starch grains appear bright. These are both simple and compound starch grains. The simple grains are circular and exhibit central dark circular spot and +–shaped dark polarimark (Figure 14.1H). The compound starch grains are large and polyhedral in shape (Figure 14.1I).

3.2.3. Powder Microscopy

The powder of the rhizome shows the following inclusions.

1. **Starch grains:** Starch grains are abundant in the powder. They are highly variable in size and shape. They are circular, semicircular, ovoid and elliptical in shape. The size of the grains ranges from 5-10 μm (Figures 14.1J, K).

2. **Vascular elements**: Xylem elements are seen as broken fragments. They have annular thickenings and close spiral thickenings. The elements are 20 μm wide (Figure 14.1L).

3. **Parenchyma cells**: Large masses of parenchyma cells are seen scattered. The cells have dense starch inclusions (Figure 14.1M).

4. **Laticifers:** Thick, darkly stained worm-like laticiferous tubes are occasionally seen in the powder. The tubes non septate, unbranched and thin walled. The laticifer is 20 μm thick (Figure 14.1N).

3.3. Physico-chemical/Fluorescence Studies

The results of physico- chemical analysis, extractive values and fluorescence characters are given in Tables 14.1 and 14.2. Preliminary phytochemical screening of the plant material was done and results are presented Table 14.3.

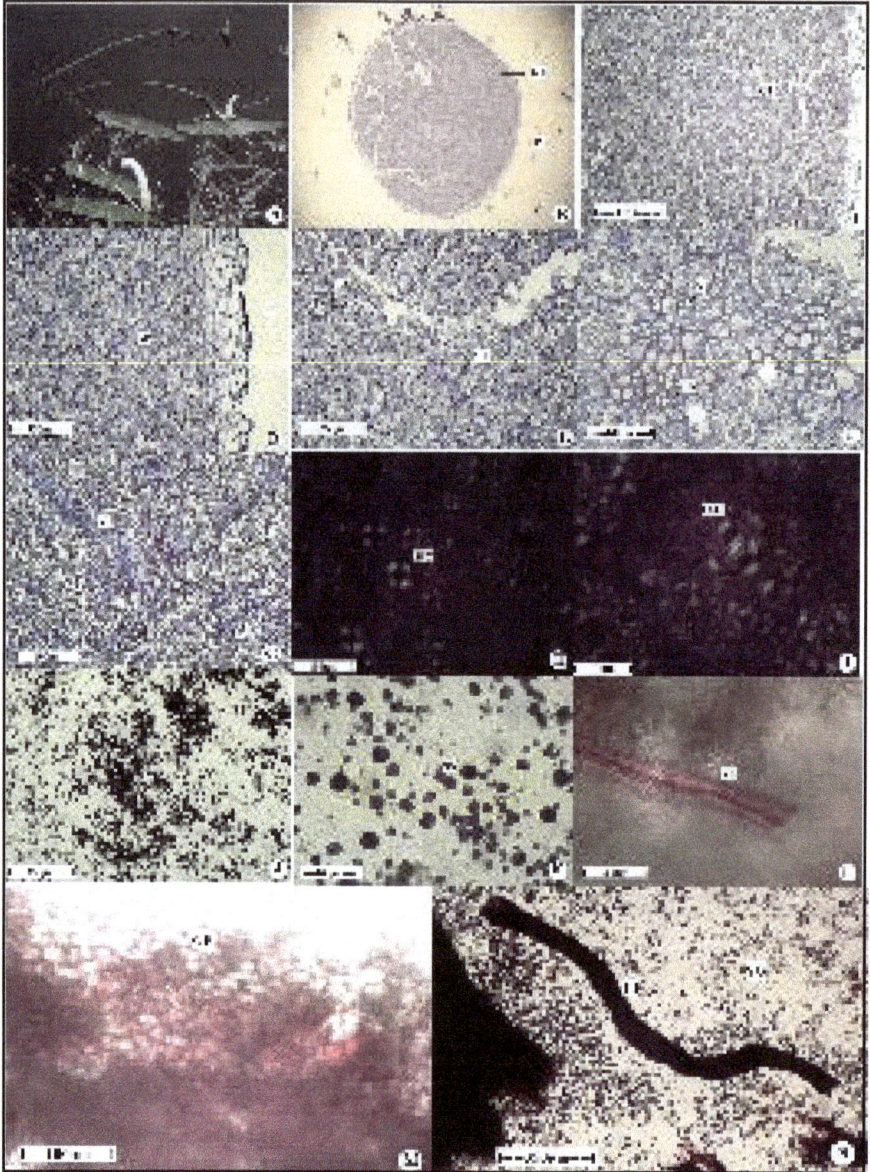

Figure 14.1A–N: A: Habit of *Aponogeton natans* L.; B: T.S. of rhizome–Entire view; C: T.S. of rhizome–A sector; D: T.S. of rhizome-periderm–enlarged; E: T.S. ground parenchyma and vascular strand; F: Vascular elements enlarged. G: Vascular strand running across the ground parenchyma; H: Simple spherical starch grains; I: Compound polyhedral starch grains; J–K: Starch grains with IKI; L: Xylem element with spiral thickenings; M: Parenchyma cells; N: Laticifer.

Abbreviations: CSG: Compound starch grains; GP: Ground parenchyma; GT: Ground tissue; Lf: Laticifer; Pe: Periderm; Ph: Phloem; SG: Starch grains; SSG: Simple starch grains; Vs: Vascular strand; X: Xylem; XE: Xylem elements.

Table 14.1: Physico-chemical characteristics of *Aponogeton natans.*

Particulars	Percentage Values
Total ash value	14.5
Water soluble ash value	12.5
Sulphated ash value	4.3
Acid insoluble ash value	2.3
Moisture content	11.6
Extractive values (Successive extraction)	
(a) Ethyl acetate	3
(b) Acetone	7
(c) Ethanol	8
(d) Water	6

Table 14.2: Fluorescence characters of *Aponogeton natans.*

Particulars of the Treatment	Day Light	254nm
Powder + Benzene	Brown	Green
Powder + Ethyl acetate	White	White
Powder + 50 per cent Aqueous ethanol	White	Green
Powder +1M HCl	Brown	Green
Powder +1N NaOH Aqueous	Brown	Green
Powder + 1N NaOH Alcoholic	Brown	Green
Powder + Acetic acid	White	Green
Powder + Nitric acid+ Ammonia	Brown	Green
Powder + Conc Nitric acid	Brown	Green
Powder + 50 per cent H_2SO_4	Brown	Brown

Table 14.3: Preliminary phytochemical screening of *Aponogeton natans.*

Phytochemicals	Ethanol	Ethyl Acetate	Acetone
Alkaloids	–	–	–
Carbohydrates	+	–	–
Amino acids	+	–	+
Phytosterols	–	+	+
Glycosides	+	–	–
Flavonoids	+	–	+
Phenols	–	–	–
Tannins	–	–	+
Saponins	+	+	+
Quinones		–	+ –

The ethanol extract showed the presence of Carbohydrates, Amino acids, Glycosides, Flavonoids, and Saponins. The ethyl acetate extract has tested positively for Phytosterols, Saponins and Quinones. Acetone extract showed the presence of Amino acids, Phytosterols, Flavonoids, Saponins and tannins. All the three extracts have Saponins compounds. These metabolites have been shown to be responsible for various therapeutic activities of medicinal plants (Trease and Evans, 1989).

4. Conclusion

Since *Aponogeton natans* has numerous uses in traditional medicine to treat several ailments such as skin diseases and wound healing, it is essential to standardize it for use as drug. The pharmacognostic constants for the rhizomes and the diagnostic microscopic features reported in this work could be further useful to evaluate pharmacological and therapeutic efficacy of this plant to claim as drug. The evaluation of pharmacognostic features should be highly useful in the preparation of herbal section of proposed pharmacopoeia.

5. Acknowledgement

Authors are sincerely thankful to University Grants Commission, New Delhi for providing financial assistance to carry out this work.

References

Anonymous, 1985. *Indian Pharmacopoeia*. Government of India, Controller of Publication, Ministry of Health and Family Welfare, New Delhi, India, p. 70.

Britto, J. and Mahesh, R., 2007. Exploration of *Kani* tribal botanical knowledge in Agasthiayamalai biosphere reserve, South India. *Ethnobotanical Leaflets*, 11: 258–265.

Esau, K., 1964. *Plant Anatomy*. John Wiley and Sons, New York, p. 767.

Gamble, J.S., 1935. *Flora of Presidency of Madras*. Botanical Survey of India, p. 337.

Harborne, J.B., 1998. *Phytochemical Methods: A Guide to Modern Techniques of Plant Analysis*, 3rd edn., pp. 1–302.

Johansen, D.A., 1940. *Plant Microtechnique*. McGraw Hill Book Company, Inc., New York, p. 523.

Kokate, C.K., Purohit, A.P. and Gokhale, S.B., 2003. *Pharmacognosy*. Nirali Prakashan, Pune, India, p. 1–624.

O'Brien, T.P., Feder, N. and McCully, M.E., 1964. Polychromatic staining of plant cell walls by Toluidine Blue O. *Protoplasma*, 59: 368–373.

Sass, J.E., 1940. *Elements of Botanical Micro-technique*. McGraw Hill Book Company, Inc., New York, p. 222.

Yoganarasimhan, S.N., 2000. *Medicinal Plants of India*. Tamil Nadu Regional Research Institute, Bangalore, India, p. 25.

Trease, G.E. and Evans, W.C., 1989. *Pharmacognosy*, 13th edn. ELBS Oxford University Press, London, p. 245–263.

Scientific Basis of Herbal Medicine (2013)
Editor: Dr. Parimelazhagan Thangaraj
Published by: DAYA PUBLISHING HOUSE, NEW DELHI

Pages 133–140

Chapter 15

In vitro Antioxidant Potential of the Leaf Extracts of *Centella asiatica* (L.) Urban, *Bacopa monniera* (L.) Pennell and *Sida acuta* Burm. F.

S. Gopika and K. Hemalatha

Department of Botany Avinashilingam University,
Coimbatore – 43, T.N., India

1. Introduction

Natural products such as herbs, fruits and vegetables become popular in recent years due to their therapeutic value, increasing interest among consumers and scientific community. A wide variety of Plant derived antioxidants have been screened *viz.*, alkaloids, phenolic acids, flavonoids (Klimezak *et al.*, 2007), terpenoids, steroids, saponins, glycosides, lignins, tannins carotenoids and vitamins (Govindarajan *et al.*, 2005). Antioxidants have become synonymous with good health. The regular consumption of dietary antioxidants may reduce the risk of several serious diseases (Gupta, 2009).Under normal conditions; free radicals generated in the body can be removed by body's natural antioxidant defences. Antioxidants can be classified into two major classes *i.e.*, enzymatic and non-enzymatic.The enzymatic antioxidants are produced endogenously and include superoxide dismutase, catalase and glutathione peroxidase. The non-enzymatic antioxidants include tocopherol, carotenoids, ascorbic acids, flavonoids and tannins which are obtained from natural plant sources (Lee *et al.*, 2004 and Gulcin *et al.*, 2003). Numerous plant derived antioxidants have already been isolated from various species. In addition, various traditionally renowned medicinal systems *viz.*, Ayurveda, Unani, Siddha and Chinese are bestowed with therapeutically active natural antioxidants (Jagannathan, 2007). In Indian system of medicine, *Centella asiatica, Bacopa monniera* and *Sida acuta* were used as a rejuvenator, tonic and as a traditional medicine for various ailments. The mechanism of pharmacological action and *in vitro* antioxidant activity of the leaves of all the three medicinal species have been investigated.

2. Materials and Methods

An investigation was carried out to analyse the *in vitro* antioxidant activity of *Centella asiatica, Bacopa monniera* and *Sida acuta*.

2.1. Plant Material

The plant samples *Centella asiatica* (L.) Urban, *Bacopa monniera* (L.) Pennell, *Sida acuta* Burm. F. were collected from Palakkad. The leaves of the three different leaf samples were shade dried and made into fine powder with homogenizer.

2.2. Preparation of Leaf Extracts

Different extracts were made from powdered leaves using petroleum ether, methanol and water. About 40 g of each leaf samples were soaked with the appropriate solvent. All the extracts were left under shaking conditions at room temperature for 24 hr. The resultant extracts were filtered using Whatman No. 1 filter paper and used for further analysis. All the analyses were done in triplicates.

2.3. *In vitro* Antioxidant Potential

2.3.1. Enzymatic Activity

☆ Estimation of Catalase activity (Luck, 1947)

☆ Estimation of Peroxidase activity (Reddy *et al.*, 1995)

☆ Estimation of Polyphenol oxidase activity (Esterbauer *et al.*, 1977)

☆ Estimation of Glutathione-s-transferase activity (Habig *et al.*, 1974)

☆ Estimation of Glutathione peroxidase (Ellman, 1959)

2.3.2. Non-Enzymatic Activity

☆ Estimation of Ascorbic acid (Roe and Keuther, 1953)

☆ Estimation of a-Tocopherol (Emmerie method 1938 -Rosenberg, 1992)

☆ Estimation of Total Polyphenol (Malick and Singh, 1980)

2.4. DPPH Radical Scavenging Assay

☆ Estimation of IC_{50} value of the methanolic leaf extracts

3. Results and discussion

3.1. Antioxidant Potential of the Methanolic Leaf Extracts of C. asiatica, B. monniera and S. acuta

The complex antioxidant system are very important for protecting cellular membranes and organells from the damaging effects of free active oxygen species. The levels of enzymatic and non-enzymatic antioxidants were assessed to represent the basic antioxidant potential of the methanolic the leaf extracts of *C. asiatica*, *B. monniera* and *S. acuta*.

3.1.1. Enzymatic Antioxidants

Antioxidants are naturally occurring nutrients or enzymes that help to maintain health. They are found in certain foods that neutralize free radicals. They include vitamin C, E, A and the minerals, copper, zinc and selenium. Perhaps the other bioactive compounds such as phytochemicals in plants is believed to have greater antioxidant effects than vitamins and minerals (Padmaja *et al.*, 2010).

Table 15.1: Estimation of enzymatic antioxidant levels of the methanolic leaf extracts of *C. asiatica, B. monniera* and *S. acuta*.

Sl.No.	Plant Name	Enzymatic Antioxidant				
		Catalase (µg/g)	Peroxidase (µg/g)	Glutathione-s-transferase (µg/g)	Glutathione peroxidase (µg/g)	Polypheno-l oxidase (µg/g)
1.	*C. asiatica*	18.08±1.2	17.34±1.97	90.70±1.73	35.67±2.37	34.87±0.03
2.	*B. monniera*	21.57±1.77	13.77±1.49	122.89±1.86	23.88±1.78	55.28±0.06
3.	*S. acuta*	39.42±2.01	21.55±1.92	145.54±1.81	25.24±1.98	97.02±0.02
	SEd	1.38	1.47	1.48	1.67	1.95
	CD (0.05)	3.39	3.61	3.7	4.1	2.23

Values are expressed by Mean±SD of three replicates.

The catalase activity in the methanolic leaf extracts ranged widely from 39.42±2.01 µg/g of protein to 18.08±1.2 µg/g protein. The maximum catalase activity was exhibited by the leaf extracts of *S. acuta* (39.42±2.01 µg/g protein) and the lowest value was observed in the leaf extract of *C. asiatica* (18.08±1.2 µg/g protein) whereas the moderate catalase activity was found to be 21.57±1.77 µg/g protein contributed by the leaf extract of *B. monniera*. In addition, Premkumar and Suriyavanthana (2010) also supported that the catalase activity of the plant extracts has antioxidant potential.

The level of peroxidase activity was moderate in the leaf extract of *C. asiatica* (17.34±1.97 µg/g protein) and *S. acuta* showed the highest value 21.55±1.92 µg/g protein for the peroxidase activity whereas the lowest level of peroxidase activity was observed in *B. monniera* (13.77±1.49 µg/g protein).

From the results obtained, it is found that ROS detoxification agents in cells includes antioxidative enzymes such as ascorbate oxidase, peroxidase, catalase. Ascrobate oxidase and peroxidase were important enzymatic antioxidants. They scavenge hydrogen peroxide as co-substrate where peroxidases are home containing enzymes capable of oxidizing organic and inorganic compounds (Murata *et al.*, 2008 and Lee *et al.*, 2007).

The results indicated that the activity of glutathione-s-transferase is found to be maximum in *S. acuta* (145±1.81 µg/g protein) compared to leaf extracts of *B. monniera* (122.89±1.86 µg/g protein) and *C. asiatica* (90.70±1.73 µg/g protein).

The leaf extracts of *B. monniera* exhibited decreased level of glutathione peroxidase activity (23.88±1.78 µg/g protein) when compared to the leaf extracts of *C. asiatica* and *S. acuta*. The leaf extract of *C. asiatica* showed lowest level of Glutathione-s-transferase activity (90.70±1.73 µg/g protein) but had maximum level of Glutathione peroxidase activity (35.61±2.31 µg/g protein) when compared to the leaf extracts of *B. monniera* and *S. acuta*. According to Mathew and Abraham (2006), the results obtained with respect glutathione-s-transferase activity have direct effect on the potential of methanolic leaf extracts to exhibit antioxidant activity.

The activity of polyphenol oxidase found to be maximum in *S. acuta* (97.02±0.02 µg/g protein) and lowest value was estimated in *C. asiatica* (34.87±0.03 µg/g of protein) and the moderate value was estimated in *B. monniera* (55.28±0.06 µg/g of protein).

3.1.2. Non Enzymatic Antioxidants

Non enzymatic antioxidant activity of leaf extracts of *C. asiatica*, *B. monniera* and *S. acuta*.

Table 15.2: Estimation of non enzymatic antioxidant levels of the methanolic leaf extracts of *C. asiatica*, *B. monniera* and *S. acuta*.

Sl.No.	Plant Name	Enzymatic Antioxidant		
		Ascorbic Acid[a]	α-tocopherol[b]	Total Polyphenol[c]
1	C. asiatica	33.2±0.11	63.41±2.15	62.3±1.15
2	B. monniera	76.4±0.69	64.31±1.52	73.3±1.20
3	S. acuta	142.3±0.61	69.63±0.94	83.3±0.96
	SEd	0.03	1.32	0.14
	CD (0.05)	0.08	3.23	0.34

Values are expressed by Mean±SD of three replicates.

a: Expressed as µg of ascorbic acid/g of sample; b: Expressed as µg of tocopherol/g of sample; c: Expressed as µg of catechol/g of sample.

It is evident that the significantly highest value of ascorbic acid (142.3±0.61 µg/g) and α-tocopherol (69.63±0.94 µg/g) and total polyphenol (83.3±0.96 µg/g) were observed in the leaf extract of *S. acuta*. Leaf extract of *C. asiatica* was found to exhibit the lowest ranges of various non enzymatic antioxidant activity *viz.*, ascorbic acid (33.2±0.11 µg/g) and α-tocopherol (63.41±2.15 µg/g) and total polyphenol (62.3±1.15 µg/g). The level of ascorbic acid, α-tocopherol and polyphenol were found to be moderate in *B. monniera*.

Maisuthiakul *et al.* (2008) revealed that plants with high antioxidant activities also have high total phenolic and flavonoids content.

3.2. *In vitro* Antioxidant Activity of the Methanolic Leaf Extracts using DPPH Assay

The antioxidant activity of the three methanolic leaf extracts were evaluated according to their ability for scavenging free radicals using DPPH assay and their results were depicted in.

Table 15.3: Estimation of antioxidant activity of the methanolic leaf extracts using *in vitro* DPPH assay.

Sl.No.	Control/Plant name	Graded Concentrations of the Leaf Samples (µg/ml)				
		20	40	60	80	100
1.	*C. asiatica*	12.23±1.98	32.54±1.95	43.94±1.49	61.79±1.64	81.01±2.28
2.	*B. monniera*	15.33±2.57	43.71±3.21	56.15±1.48	65.87±1.34	87.43±3.69
3.	*S. acuta*	27.55±3.76	46.65±3.16	67.75±2.15	74.56±1.60	95.29±2.83
4.	Standard (Ascorbic acid)	17.10±.56	45.20±2.40	69.25±3.11	73.20±1.99	92.29±2.40
	SEd	0.24	0.26	1.77	1.35	2.33
	CD (0.05)	0.55	0.6	4.08	3.13	5.38

Values are expressed by Mean±SD of three replicates.

Among the three extracts of *C. asiatica*, *B. monniera* and *S. acuta* and standard tested for the *in vitro* antioxidants potential, the crude methanolic extracts of *S. acuta* (95.29±2.83 per cent) showed strongest DPPH scavenging activity at 100 µg/ml compared to *B. monniera* (87.43±3.69 per cent) and *C. asiatica* (81.01±2.28 per cent) when compared with standard (92.29±2.40 per cent). A dose-dependent relationship was found in the DPPH scavenging activity. The activity increased with an increase in the dose of each methanolic leaf extracts. At graded concentrations of (20 to 100 µg/ml), *S. acuta* had free radical scavenging activity from 27.55±3.76 per cent to 92.29±2.83 per cent. *Bacopa monniera* and *Centella asiatica* had the lowest percentage of activity compared to the standard. The percentage of inhibition by the standard (ascorbic acid) (92.29±2.83 per cent) was found to be higher compared to methanolic extracts of *Bacopa monniera* and *Centella asiatica*.

DPPH assay is one of the most widely used methods for screening antioxidant activity of plant extracts (Lee *et al.*, 2004; Mensor *et al.*, 2001; Yildrim *et al.*, 2001 and Huang and Kuo, 2000).

Mathew and Abraham (2006) stated that the increasing levels of enzymatic and non enzymatic antioxidants were directly correlated with the antioxidant activity.

3.2.1 Comparison of IC_{50} Values and Antioxidant Activity of the Methanolic Leaf Extracts

IC_{50} values (concentration of sample required to scavenge 50 per cent free radical or to prevent lipid peroxidation by 50 per cent) were calculated from the regression equations. IC_{50} values is inversely proportional to the antioxidant activity. Higher the IC_{50} values lower the ability of the methanolic extracts to scavenge hydroxyl radicals (Dasgupta and De, 2005). The results showed that the comparative data of DPPH activity, as determined by the IC_{50} of the three methanolic extracts of *C. asiatica*, *B. monniera* and *S. acuta*. Among the leaf extracts, *Sida acuta* showed highest activity with low IC_{50} value (13.85 µg/ml). Lowest antioxidant activity was observed in *C. asiatica* with high IC_{50} value (19.20 µg/ml). The highest antioxidant activity of the methanolic leaf extract of *S. acuta* depends upon the scavenging of free radical as reactive oxygen species are more reactive and induces severe damages to adjacent biomolecules.

Table 15.4: Comparison of DPPH radical scavenging activity and IC_{50} values of *C. asiatica*, *B. monniera* and *S. acuta*.

Sl.No.	Control/ Plant Extracts	IC_{50} (µg/ml)
1.	C. asiatica	19.20
2.	B. monniera	16.40
3.	S. acuta	13.85
4.	Standard (Ascorbic acid)	18.10

It is evident from the study that DPPH scavenging is widely used to test the free radical scavenging activity of several natural products (Ahn *et al.*, 2007). DPPH is a stable free radical and any molecule that can donate an electron or hydrogen atom to DPPH and can react with it which bleach the DPPH absorption at 517 nm (Huang *et al.*, 2005). But there is a reverse correlation between IC_{50} value and DPPH scavenging activity.

4. Conclusion

The present study showed a rich array of potentially active compounds. Among the three leaf extracts *S. acuta* showed highest antioxidant activity compared to *B. monniera* and *C. asiatica*.

References

Ahn, R., Kumazawa, S., Clsui, Y., Nakaura, J., Matsuka, M., Zhu, F. and Nakayam-a, T., 2007. Antioxidant activity and constituents of propolis collected in various areas of China. *Food Chem.*, 101: 1383–1392.

Dasgupta, N. and Bratati, De, 2005. Antioxidant activity of some leafy vegetable of India: A comparative study. *J. Food Chem.*, 10: 417–474.

Ellman, G., 1959. Tissue sulphydryl groups. *Archiv. of Biochem. and Biophy.*, 32: 70–77.

Esterbauer, H., Schwarzyl, E. and Hayn, M., 1977. A rapid assay for catechol and laccase using 2-nitro-5-thio-benzoic acid. *Anal. Biochem.*, 77: 486–497.

Govindarajan, R., Vijayakumar, M. and Pushpangadan, P., 2005. Antioxidant approach to disease management and the role of 'Rasayana' herbs of ayurveda, *Journal of Ethnopharmacology*, 99: 165–178.

Gulcin, I., Kufrevioglu, O.I., Oktay, M. and Buyukokuroglu, M.E., 2003. Antioxidant, antimicrobial, antiulcer and analgesic activities of *Uteica dioca*. *J. Ethno-Pharma.*, 90: 205–215.

Gupta, S. and Prakash, J., 2009. Studies on Indian green leafy vegetables for their antioxidant activity. *Plant Foods for Human Nutrition*, pp. 61–64

Habig, W.H., Pabst, M.J. and Jacopy, W.B., 1974. Glutathione s-transferase: The first enzymatic step in mercapturic acid formation. *J. Biol. Chem.*, 249: 7310–7339.

Huang, D., Ou, B. and Prior, R.L., 2005. The chemistry behind the antioxidant capacity assay. *J. Agric. Food Chem.*, 5: 1841–1856.

Jagannathan, R., 2007. CMO tag suits. *Ind. Biotech. Advanced Biotech.*, 3–25.

Klimezak, I., Nalecka, M., Szlachta, M. and Glizezyns-ka-Swiglo, A., 2007. Effect of storage on the content of polyphenols, vitamin C and the antioxidant activity of orange juices. *J. Food Compos. Anal.*, 20: 313–322.

Lee, J., Koo, N. and Min, D.B., 2004. Reactive oxygen species, ageing and antioxidative nutraceuticals. *CRFSFS*, 3: 21–33.

Lee, S.H., Ashan, N., Lee, K.W., Kim, D.H., Lee, D.G., Kwak, S.S., Kwon, S.Y., Kim, T.H. and Lee, B.H., 2007. Simultaneous over expression of both Cu, Zn superoxide dismutase and ascorbate peroxidase in transgenic tall fescue plants confers increased tolerance to a wide range of abiotic stresses. *J. Plant Physiol.*, 164: 1626–163.

Luck, H., 1947. Catalase. In: *Methods in Enzymatic Analysis*, (Ed.) Biogneyer. Academic Press, New York, pp. 85–88.

Maisuthisakul, P., Pasut, S. and Ritthirvangdej, P., 2008. Relationship between antioxidant properties and chemical composition of some Thai plants. *J. Food Compos. Anal.*, 21: 219–228.

Malick, C.P. and Singh, M.B., 1980. *Plant Enzymology and Histoenzymology*. Kalyani Publications, New Delhi, pp. 286–289.

Mathew, S. and Abraham, E.T., 2006. *In vitro* antioxidant activity and scavenging activity and scavenging effects of *Cunnamomum verum* laf extract assayed by different methodologies. *Food Chem. Toxicol.*, 44: 198–206.

Mensor, L., Menzes, L.F.S., Leitao, A.S., Rels, A.S. and Santos, T.C., 2001. Screening of Brazaile plant extracts. *Food Chem.*, 112: 595–598.

Murata, K., Nakamura, N. and Ohno, H., 2008. Affecting the oxidative capacity of acre oxidase by an electro chemical approach. *Commun.*, 367: 457–461.

Padmaja, M., Sravanthi, M. and Hemalatha, K.P.J., 2010. Evaluation of antioxidant activity of two Indian medicinal plants. *Journal of Phytology*, 3(3): 86–91.

Premkumar, P., Priya, J. and Suriyavathan, M., 2010. Evaluation of antioxidant potential of *Andrographis echioides* and *Boerhaavia diffusa*. *International Journal of Current Research*, 3: 59–62.

Reddy, K.P., Subhani, S.M., Khan, P.A. and Kumar, K.B., 1995. Effect of light and benzyl adenine on dark treated growing rice (*Oryza sativa*) leaves, changes in peroxidative activity. *Plant Cell Physiol.*, 26: 987–944.

Roe, J.H. and Keuther, C.E., 1953. The determination of ascorbic acid in whole blood and wine through 2, 4-dinitro phenyl hydrazine derivative of dehydro ascorbic acid. *J. Biol. Chem.*, 147: 399–407.

Rosenberg, H.R., 1992. *Chemistry and Physiology of the Vitamins.* Interscience Publishers Inc., New York, pp. 452–543.

Yildrim, M., Oktay and Bilaloglue, V., 2001. The antioxidant activity of the leaves of *Cydonia vulgaris. Turkist J. Med. Sc.,* 31: 23–27.

Scientific Basis of Herbal Medicine (2013)
Editor: Dr. Parimelazhagan Thangaraj
Published by: DAYA PUBLISHING HOUSE, NEW DELHI

Pages 141–148

Chapter 16

Micropropagation of Shoot-Tip Explants of *Aerva lanata* (L.) A.L. Juss. Ex Schultes.: A Resourceful Medicinal Plant

Desingu Kamalanathan and Devarajan Natarajan

Natural Drug Research Laboratory,
Department of Biotechnology, Periyar University,
Salem – 636011, Tamil Nadu, India

1. Introduction

Aerva lanata (L.) A. L. Juss. ex Schultes (locally called as 'Poolapoo') is an erect or prostrate under-shrub with a long tap-root and many wolly-tomentose branches (Bhattacharjee, 2004) belongs to the family Amaranthaceae. It is also known as Chaya (Hindi) and Bhadra (Sanskrit). It is distributed in the wild, throughout India, Arabia, Tropical Africa, Sri Lanka, Phillipine and Java (Krishnamurthi, 2003).

The plant is used in arresting haemorrhage during pregnancy, to dissolve kidney and gallbladder stones, for uterus clearance after delivery and to prevent lactation (Yoga narasimhan *et al.*, 1979; Upadhay *et al.*, 1998; Singh *et al.*, 1980; Chetty *et al.*, 1989; Vedhavathi *et al.*, 1990; Sudhakar *et al.*, 1998 and John 1984). In traditional medicine, the plant is used in fractures, nasal bleeding, cough, strangury (slow to be and painful discharge of urine), headache, spermatorrhoea and urolithiasis (Mukergee *et al.*, 1984; Sikarwar *et al.*, 1993 and Girach *et al.*, 1994).

The pharmacological aspects of the plant reported to contain diuretic, anti-inflammatory, hypoglycemic, anti-diabetic, antiparasitic, hepatoprotective, anti-urolithiasis, antiasthmatic, antifertility and hypolipidemic properties (Vetrichelvan *et al.*, 2000; Dulaly, 2002; Pullaiah, 2003; Shirwaikar *et al.*, 2004). The nature of phytochemical constituents consists of sitosteryl palmitate, hentriacontane, betulin etc were isolated from this plant (Chandra and Sastry 1990, Aiyar *et al.*, 1973). The major alkaloids *i.e.* Ervine, methylervine, ervoside, aervine, methylaervine, aervoside, ervolanine, and aervolanine were isolated and flavanoids such as Kaempferol, quercetin, isorhamnetin, persinol, persinosides A and B, and other phyto-constituents such as methyl grevillate, lupeol, lupeol acetate benzoic acid, β-sitosteryl acetate and tannic acid (Afaq *et al.*, 1991; Yuldashev *et al.*, 2002; Zapesochnaya *et al.*, 1992) etc. were reported to be present in the *A. lanata*.

A large-scale and continuous utilization of the available plant materials has leads to become a serious issue to loss of population and diversity in the ecosystem near future. So keeping these medicinal importance and necessity in mass development of the raw materials from the plants required a biological alternate to utilize the source without interruption and exploitation, hence the tissue culture techniques will sort-out the requirements using few segment of explants. The current work was aimed to develop an efficient micro propagation protocol for *Aerva lanata* (L.) A.L. Juss. ex Schultes using shoot-tip explant, and to conserve the plant under natural environment.

2. Materials and Methods

2.1. Collection of Plant Materials

The elite mother plants were collected from Botanical garden, Periyar University premises, Salem and it was botanically identified by Dr. D. Natarajan, Assistant Professor, Department of Biotechnology, Periyar University, Salem. The voucher specimen was deposited in the Natural Drug Research Laboratory, Department of Biotechnology, Periyar University, Salem. The collected plants were finely trimmed and shoot tip explants were chosen for inoculation.

2.2. Media Preparation

Modified MS media (Murashige and Skoog, 1962) was prepared and supplemented with 3 per cent sucrose as carbon source and 0.7 per cent agar (Hi Media) along with various concentrations of cytokinins and auxins, for the shoot propagation and root proliferations with BAP alone (Basal, 1.0, 1.5, 2.0, 2.5 mg/l) and IBA alone and in combination with IAA and NAA (1.5, 2.0 IBA, 0.5 + 0.5, 0.5 + 1.0 IBA + IAA and NAA respectively). The pH of the media was adjusted to 5.8 using 0.1N HCL and 0.1N NAOH prior to autoclaving for 15 min at 121p C. The prepared medium was cooled and stored for inoculation purposes.

2.3. Surface Sterilization and Inoculation

The explants (shoot tips) were excised aseptically with sterile scissors and washed with running tap water (for 2 – 5 min) followed by treatment with 0.5 per cent (w/v) Bavistin (fungicide for 2 min) and rinsed thoroughly with sterile distilled water, then surface sterilized initially with 70 per cent ethanol for 30 seconds followed by treatment with 0.5 per cent (v/v) Sodium hypochlorite (for 2 min) and explants were then surface disinfected with 0.1 per cent $HgCl_2$ (w/v) solution (for 3 – 10 min) after each treatments the explants were rinsed thoroughly with sterile distilled water (3–5 times), and finally trimmed with sterile blade and wet explants are dried in No. 1 whatmann filter paper were inoculated aseptically and kept for incubation undisturbed.

2.4. Incubation and Multiple Shoot Formation

Inoculated cultures were maintained under cool-white fluorescent light at 24±2°C with 16 h photoperiod. The shoot tips were continuously monitored for contamination and regeneration of the inoculated plants and maintained for 3-4 weeks periods. Then, the auxiliary buds were isolated and repeatedly sub cultured on same medium for the formation of multiple shoots and *in vitro* flowering for next 4 weeks periods. Then, the well developed shoots were transfer to *in vitro* rooting.

2.5. Rooting and Acclimatization

For root formation, well developed shoots (5 cm) were transferred to MS media supplemented with auxins (IBA) alone and combination with NAA (0.5 – 2.0 IBA, 0.5 + 0.25 – 1.0 mg L⁻¹ NAA, IBA + IAA 0.5 – 1.0 mg L⁻¹ respectively) and incubated as same conditions. Each treatment consists of 3 replications and each replication, 21 explants were used and the data was recorded after 5 weeks of culture. The well developed plantlets were transferred to polycups containing sterile soil and sand (1: 1 ratio) and watered often and transferred to green house.

3. Results and Discussion

3.1. Shoot Induction

The hormonal strength (MS medium) exhibits the potential differences in the regeneration of inoculated explants. The MS basal medium without any PGR's showed slight difference in the regeneration of the inoculated explants produced shoots upto 1.2±0.3 cm length, whereas the explants inoculated in the MS medium supplemented with different concentration of hormone 6- Benzyl adenine/Benzyl Amino Purine (BAP) showing more percentage of regeneration (average of 70 – 80 per cent). The highest shoot length (about 5.7±0.8 cm) was observed in the concentration of 1.0 mg/L BAP and the mean of 82 per cent of multiple shoots. The rate of regeneration of the inoculated explants is 89 per cent.

Whereas, other concentrations are considerably near to highest shoot formation and regeneration but the elongation of shoot length were quite lower in the concentration of 1.5 and 2.5 mg/L BAP showing 80 per cent are regeneration with the mean of 80 and 76 per cent of multiple shoots of shoot length about 4.0±0.6cm and 3.8±0.5 cm respectively (Table 16.1). The explants were maintained for 4 weeks periods without any disturbance. Similar reports were shown by several authors the role of cytokinins used for enhancement and successful regeneration of explants in legumes by Malik and Saxena 1992; Murthy *et al.*, 1995; Cruz de Carvalho *et al.*, 2000. A few other reports describing the importance of hormonal regulations in development of tissue culture medium in *Amaranthus edulis* and *Alternanthera sessils* were reported by Bui van *et al.*, 1998 and Alveera Singh *et al.*, 2009. Similarly, this is the first hand information on hormonal based regeneration of *A. lanata* by shoot tip explants *in vitro*.

3.2. Multiple Shoots Formation and *In vitro* Flowering

The mature cultures were isolated and it was repeatedly sub cultured in the new medium at same hormonal concentration produced highest rates of multiple shoots and they were continuously maintained for 4 weeks. The regenerated plants produced *in vitro* flowerings in the medium containing 1.0 mg/L BAP. The similar report was observed by Chang and Chang (2003) in *Cymbidium ensifolium;* Wang *et al.* (2001) in bitter melon and Nistch and Nistch (1967) in *Plumbago indica.*

Table 16.1: Regeneration and multiple shoots formation of different concentration of hormones.

Concentration of Hormones (mg/L)	Mean of the Shoot Length (cm)	Rate of Regeneration (per cent)	Mean Number of Multiple Shoots (per cent)
MS basal	1.2±0.3	65	46
1.0 BAP	5.7±0.8	89	82
1.5 BAP	4.0±0.6	80	80
2.0 BAP	3.4±0.6	80	74
2.5 BAP	3.8±0.5	80	76

Mean = the average of 14 explants in triplicates±SE.

3.3. *In vitro* Rooting

The well grown *in vitro* shoots (upto 5 cm length) were dissected after 12 weeks maturation and were inoculated in the MS medium supplemented with auxins IBA alone, combination of with NAA and IAA solidified with 0.6 per cent agar. The roots were initiated after 10 – 15 days of incubation in the cut ends of the shoots. The formation of main and branch roots were higher in MS medium fortified with 2.0 mg/L IBA. The mean of root length was about 7.2±0.3 cm and the percentage of branch roots about 79 per cent and other concentration produced comparable rates of root proliferation. Whereas, the combination of IBA with NAA and IAA produced lower rate of roots and the average length of roots was about 3.8±0.6 cm in IBA + IAA and 3.3±0 in the IBA + NAA (Table 16.2). Our results were comparable with the reports of Wesely *et al.* (2012) in *A. aspera,* Saha *et al.* (2010) in *Coleus forskohlii,* and Subramanian *et al.* (2011) in *Vicoa indica* reported the IBA induced rooting in the MS medium.

Table 16.2: Root formation of various concentrations of hormones.

Hormones	Concentration of Hormones (mg/L)	Mean of the Root Length (cm)	Mean of the Branching Roots (per cent)
IBA	1.5	5.8±0.1	68
	2.0	7.2±0.3	79
IBA + IAA	0.5 + 0.5	3.8±0.6	56
	0.5 + 1.0	3.6±0.2	54
IBA + NAA	0.5 + 0.5	3.3±0	52
	0.5 + 1.0	2.8±0.4	40

Mean = the average of 14 explants in triplicates±SE.

Figure 16.1: Mass propagation of *A. lanata* using Shoot tip explants via direct organogenesis.

A: Multiple shoots (shoot-tip); B: *In vitro* flowers; C: *In vitro* rooting; D: Acclimatized plant.

3.4. Acclimatization

After completion of root formation (after 6 weeks of inoculation) the plants were successfully acclimatized by transferring the plants with complete shoots, flowers and roots in a pot mixture of sterile sand and red soil in 1: 1 ratio. The rooted plants were carefully removed and washed with sterile water and planted in the polycups containing sand and soil and watered with ½ MS medium frequently and kept at 28°C for few weeks and these plants were transferred to garden. The average rate of successful acclimatization was about 78 per cent.

4. Conclusion

The present study was bringing with an efficient reproducible protocol for the regeneration of multipurpose plant (*Aerva lanata*) in short duration. This work also observed/suggests that the use of lower concentration of plant growth hormones like cytokinins and auxins helped for high regeneration rates.

Acknowledgement

The authors are gratefully acknowledged to Department of Biotechnology, Periyar University, Salem for providing excellent laboratory facilities to carry out this work successfully.

References

Afaq, S.H., Tajuddin, Afridi, R. and Bisehri Booti, 1991. *Aerva lanata* Juss.: Some lesser known uses and pharmacognosy. *Ethanobot.*, 3: 37–40.

Aiyar, V.N., Narayanan, V., Seshadri, T.R. and Vydeeswaran, S., 1973. Chemical composition of some Indian medicinal plants. *Indian J. Chem.*, 11: 89–90.

Bhattacharjee, S.K., 2004. *A Handbook of Medicinal Herbs*, 4th edn. Pointer Publishers, p. 17.

Bui van, Le, Do, M.Y, N.T., Jeanneau, M., Sadik, S. and Shanjun, T., 1998. Rapid plant regeneration of a C4 dicot species: *Amaranthus Edulis*. *Plant Sci.*, 132: 45–54.

Chandra, S. and Sastry, M.S., 1990. Chemical constituents of *Aerva lanata*. *Fitoterapia*, 61: 188.

Chang, C. and Chang, W.C., 2003. Cytokinins promotion of flowering in *Cymbidium ensifolium* var. *misericors in vitro*. *Plant Growth Regu.*, 39: 217–221.

Chetty, K.M. and Rao, K.N., 1989. Ethanobotany of Sarakallu and adjacent areas of Chittoore District Andhra Pradesh. *Vetegos.* 2: 51–58.

Cruz, de, Carvalho, M.H., Van Le, B., Zuily-Fodil, Y., Pham-Thi, A. and Tran Thanh Van, K., 2000. Efficient whole plant regeneration of common bean (*Phaseolus vulgaris* L.) using thin cell layer culture and silver nitrate. *Plant Sci.*, 159: 223–232.

Dulaly, C., 2002. Antimicrobial activity and cytotoxicity of *Aerva lanata*. *Fitoterapia*, 73: 92–94.

Girach, R.D., Aminuddin, S.P.A. and Khan, S.A., 1994. Traditional plant remedies among the kondh of district Dhenkal Orissa. *Int. J. Pharmacol.*, 32: 274–283.

John, D., 1984. One hundred useful raw drugs of the Kani tribes of Trivandrum forest division, Kerala, India. *Int. J. Crude Drug Res.*, 22: 17–39.

Krishnamurthi, A., 2003. *The Wealth of India*, Vol. 1. CSIR, New Delhi, p. 92.

Malik, K.A. and Saxena, P.K., 1992. Regeneration in *Phaseolus vulgaris* L.: High-frequency induction of direct shoot formation in intact seedling by N6-benzylaminopurine and thidiazuron. *Planta*, 186: 384–389.

Mukerjee, T., Bhalla, N., Singh, A.G. and Jain, H.C., 1984. Herbal drugs for urinary stones. *Indian Drugs*, 21: 224–228.

Murashige, T. and Skoog, F., 1962. A revised medium for rapid growth and bioassays with tobacco tissue cultures. *Physiol Plant*, 15: 473–497.

Murthy, B.N.S., Murch, S. and Saxena, P.K., 1995. Thidiazuron-induced somatic embryogenesis in intact seedling of peanut (*Arachis hypogaea*): Endogenous growth regulator levels and significance of cotyledons. *Physiol Plant.*, 94: 268–276.

Nitsch, C. and Nitsch, J.P., 1967. The induction flowering *in vitro* in stem segments of *Plumbago indica* II. The production of reproductive buds. *Planta*, 72: 371–384.

Pullaiah, T. and Naidu, C.K., 2003. *Antidiabetic Plants in India and Herbal Based Antidiabetic Research*. Regency Publications, New Delhi, pp. 68–69.

Saha, S., Haque, R., Belsare, D.P. and Bera, T., 2010. Micropropagation of an important medicinal plant *Coleus forskohlii*. *Int. J. Pharm. Sci.*, 2(1): 429–435.

Shirwaikar, A., Issac, D. and Malini, S., 2004. Effect of *Aerva lanata* on cisplatin and gentamycin models of acute renal failure. *J, Ethno-Pharmacol.*, 90: 81–86.

Sikarwar, R.L.S. and Kaushik, J.P., 1993. *Folk Medicine of the Morena District, Madhya Pradesh, India*, 31: 283–287.

Singh, Alveera, Kandasamy, Thangaraj and Odhav, Bharti, 2009. *In vitro* propagation of *Alternanthera sessilis* (sessile joyweed), a famine food plant. *Afr. J. Biotech*, 8(21): 5691–5695.

Singh, V. and Pandey, R.P., 1980. Medicinal plantlore of tribal's of Eastern Rajasthan India. *J. Econ. Taxon. Bot.*, 1: 137–147.

Subramanian, S.S., Monzhi, K.R., Chandar, B.C., Sundar, A.N. and Devi, C.M., 2011. Rooting experiments on stem cuttings of *Vicoa indica* (L.) DC.: An important medicinal plant. *J. Biosci. Res.*, 2(1): 35–37.

Sudakar, A. and Chetty, K.M., 1998. Medicinal importance of some angiospermic weeds used by the rural people of Chittoore District of Andhra Pradesh, India. *Fitoterpia*, 69: 390–400.

Upadhay, O.P., Kumar, K. and Tiwari, R.K., 1998. Ethanobotanical study of skin treatment uses in medicinal plants of Bihar. *Pharmaceutical Biol.*, 36: 167–172.

Vedhavathi, S. and Rao, K.N., 1990. Nephro protectors folk medicine of Royalaseema Andhra Pradesh. *Ancient Sci. Life*, 9: 164–167.

Vertichelvan, T., Jegadeesan, M., Senthil Palaniappan, S., Murali, N.P. and Sasikumar, K., 2000. Diuretic and anti-inflammatory activities of *Aerva lanata* in rats. *Indian J. Pharm. Sci.*, 62: 300–302.

Wang, S., Tang, L. and Chen, F., 2001. *In vitro* flowering of bitter melon. *Plant Cell Reports*, 20: 393–397.

Wesely, Edward, Gnanaraj, Johnson, Marimuthu, Mohanamathi, R.B. and Kavitha Marappampalyam, Subramanian, 2012. *In vitro* clonal propagation of *Achyranthes aspera* L. and *Achyranthes bidentata* Blume using nodal explants. *Asian Pacific Journal of Tropical Biomedicine*, 11: 1–5.

Yoga Narasimhan, S.N., Bhat, A.V. and Togunashi, V.S., 1979. Medicinal plants from Mysore District, Karnataka. *Indian Drug Pharmaceut. Ind.*, 14: 7–22.

Yuldashev, A.A., Yuldashev, M.P. and Abdullabekova, V.N., 2002. Components of *Aerva lanata. Chem. Nat. Comp.*, 38(3): 293–294.

Zapesochnaya, G.G., Kurkin, V.A., Okhanov, V.V. and Miroshnikov, A.I. (1992). Canthin-6-one and β carboline alkaloids from *Aerva lanata. Planta Med.*, 58(2): 192–196.

Scientific Basis of Herbal Medicine (2013)
Editor: **Dr. Parimelazhagan Thangaraj**
Published by: **DAYA PUBLISHING HOUSE, NEW DELHI**

Pages **149–161**

Chapter 17

Screening of Commonly Consumed Green Leafy Vegetables for Antioxidant Activity for Improvising Human Health

A. Manojkumar, B. Senthil Kumar, K. Shanmuga Priya, N. Saraswathy and P. Ramalingam

Department of Biotechnology, Kumaraguru College of Technology, Coimbatore – 641 049, Tamil Nadu, India

1. Introduction

A free radical is defined as an atom or group of atoms that have one or more unpaired electrons. Any free radical involving oxygen can be referred to as reactive oxygen species (ROS) (Bandyopathyay *et al.*, 1999). The human body has several mechanisms to counteract damage by free radicals and other reactive oxygen species. These act on different oxidants as well as in different cellular compartments. The best line of defense against free radical damage is the presence of antioxidants.

An antioxidant is a molecule stable enough to donate an electron to a rampaging free radical and neutralize it, thus reducing its capacity to damage. The best known antioxidants are vitamin E, vitamin C and the carotenoids. Antioxidants tend to neutralize free radicals before they create harmful problems to our bodies. In general, water-soluble antioxidants react with oxidants in the cell cytosol and the blood plasma, while lipid-soluble antioxidants protect cell membranes from lipid peroxidation (Hiner *et al.*, 2002). Antioxidants are capable of stabilizing, or deactivating free radicals before they attack the cells (Halliwell and Gutterudge, 1989).

The various epidemiological studies have consistently shown thatconsumption of fruits and vegetables has been associated with reduced risk of chronic diseases, such as cardiovascular diseases and cancer (Sargeant *et al.*, 2001) and neurodegenerative disorders, including Parkinson's diseases (Gella and Durany,

2009). The oxidative products, mainly the superoxide anion radical (O_2^-), from diabetic monocytes during oxidative stress lead researchers to give more attention to the protective functions of naturally occurring antioxidants (Miller *et al.*, 1997).

Fresh vegetables are the best sources of antioxidants as they contain a number of vitamins and minerals. Fruits and vegetables are packed with powerful antioxidants that can lower the risk of heart disease, cancer, diabetes-related damage and even slow down the body's natural aging process. Vegetables provide the body with an added source of antioxidants that is needed to properly wage war against free radicals. Without the necessary intake of vegetables, free radicals can spread and eventually lead to stroke, heart attack, arthritis, vision problems, Parkinson's disease, Alzheimer's disease and various types of cancer (Demple and Harrison., 1994). Plants produce a very impressive array of antioxidant compounds that includes carotenoids, flavonoids, cinnamic acids, benzoic acids, folic acid, ascorbic acid, tocopherols and tocotrienols to prevent oxidation of the susceptible substrate (Hollman, 2001). Common antioxidants include vitamin A, vitamin C, vitamin E, and certain compounds called carotenoids (such as lutein and beta-carotene) (Hayek, 2000). These plant-based dietary antioxidants are believed to have an important role in the maintenance of human health because our endogenous antioxidants provide insufficient protection against the constant and unavoidable challenge of reactive oxygen species (Holt *et al.*, 2009).

Green leafy vegetables occupy an important place among the food crops as these provide adequate amounts of many vitamins and minerals for humans. They are rich source of carotene, ascorbic acid, riboflavin, folic acid and minerals like calcium, iron and phosphorous. In nature, there are many underutilized green leady vegetables of promising nutritive value, which can nourish the ever increasing human population (Gupta and Prakash, 2009). Many of them are resilient, adaptive and tolerant to adverse climatic conditions. The nutritional value of green leafy vegetables and the lack of awareness of the antioxidant potential of some unexploited green leafy varieties have provoked us to screen the antioxidant activity of four under exploited green leafy vegetables such as *Amaranthus gangetics, Basella alba, Sesbania grandiflora, Solanum nigrum.*

2. Materials and Methods

2.1. Chemicals Used

1,1-Diphenyl-2-Picryl Hydrazyl (DPPH), methanol, ethanol, ferric chloride, phosphomolybdic acid, ammonia, chloroform, sulfuric acid, sodium hydroxide, hydrochloric acid, metallic zinc and magnesium, acetic acid, lead acetate, ammonium chloride, iodine, potassium iodide, pyridine, sodium nitroprusside, sodium bicarbonate, mercury, nitric acid, cupric acetate, potassium sodium tartarate, olive oil, ammonium molybdate, disodium hydrogen phosphate, sodium di hydrogen phosphate, potassium ferric cyanide, trichloro acetic acid (TCA), SDS, tannic acid, rutin, sodium nitrite, aluminium chloride from Himedia, Mumbai (India). All reagents used were of the analytical grade.

2.2. Plant Materials

The plant materials were collected from local market in Coimbatore (Tamil Nadu, India) during the months of December 2011. The name of the plants and plant parts used for the study are tabulated in Table 2.1

Table 2.1. Name of the plant and plant parts used.

Sl.No.	Botanical Name	Tamil Name	Plant Part Used
1.	*Amaranthus gangetics*	Mulaikeerai	Leaves
2.	*Basella alba*	Pasalaikeerai	Leaves
3.	*Sloanum nigrum*	Manathakalikeerai	Leaves
4.	*Sesbania grandiflora*	Agathikeerai	Leaves

2.3. Preparation of Plant Extracts

The leaves were shade-dried for around 5 days. The dried leaves were then crushed to made into coarse powder. Then two kinds of extraction process carried out.

2.3.1. Aqueous Extraction

Five grams of dried powder was extracted separately with 100 ml water. The extracts were boiled for 1 hour in boiling mantle. Then the contents were filtered using whatmann filter paper No. 1, onto Petri plates and extracts were evaporated at 40°C for 2 days in hot air oven. The dried extracts were scraped by adding 3 ml of water and stored at 4°C for further phytochemical and pharmacological screening.

2.3.2. Solvent Extraction

Five grams of dried powder was extracted separately with 50 ml ethanol. The extracts were placed in shaker for 2 days in conical flasks and the contents were filtered using whatmann filter paper No.1, onto Petri plates. Extracted filtrates in the Petriplates were evaporated at room temperature for 2 days. The dried extracts were scraped it by addition of 3 ml of methanol and stored at 4°C for further phytochemical and pharmacological screening.

2.4. Preliminary Phytochemical Analysis

Qualitative tests for the presence of plant secondary metabolites such as carbohydrates, alkaloids, tannins, flavonoids, saponins and glycosides and terpenoids, were carried out on the leaf powder using standard procedures (Chitravadivu *et al.*, 2009). The various phytochemical tests such as Ferric chloride test, Phosphomolybdic test, Salkowski test, Decolourization test, Ammonia test, Lead acetate test, Aluminium chloride test, Braemer's test, Dragendorff's test, Wagner's test, Keller – Killani test, Legal's test, Froth test, Sodium bicarbonate test, Terpenoids test, Millon's test, Bradford's test, Fehling's test, starch test and cellulose test were performed.

2.5. Estimation of Total Phenols by Folin-Ciocalteau method (Singleton and Rossi, 1965)

The amount of total phenolic in plant extracts was determined with Folin–Ciocalteu reagent according to the method of Singleton and Rossi (1965). One gram of the sample was weighed and ground with 10 ml of 80 per cent aqueous ethanol in a mortar and pestle. It was centrifuged at 6000 rpm for 10 minutes. Supernatant was obtained in a separate tube and pellet was re-extracted with 5 ml ethanol. Contents were centrifuged again at 6000 rpm for 10 minutes to obtain supernatants. Supernatants thus obtained were pooled and boiled to evaporate ethanol and the residue was dissolved in distilled water.

A graph was drawn by plotting the concentration of total phenolics along the X-axis and the optical density reading along Y-axis. The total phenolic was calculated using standard phenol calibration curve and the results were expressed in mg of tannic acid equivalents/g dry weight of extract.

2.6. Estimation of Flavonoids by Aluminium Chloride Method (Zhishen, Mengcheng, and Jianming, 1999)

One gram of the sample was weighed and ground with 10 ml of 80 per cent aqueous methanol in a mortar and pestle. Ground samples were filtered with the help of Whatmann filter paper No. 1 and the filtrate thus obtained to be used for further experiments.

0.5, 1.0, 1.5, 2.0 and 2.5 ml (concentration varying from 50 to 250µg) of the standard solution was pipetted out into a series of test tubes. 0.1 ml of the sample was pipetted out into a test tube. To all the tubes, including the blank, distilled water was added to make up to 2.5 ml. To all the tubes, 75 µl of 5 per cent sodium nitrite was added and incubated at room temperature for 5 minutes. Then 150 µl of 10 per cent aluminium chloride was added and incubated at room temperature for 6 minutes. Then 0.5 ml of 1M sodium hydroxide was added, mixed well and the pink colour formed was spectrophotometrically measured at 510 nm. A graph was drawn by plotting the concentration of rutin along the X-axis and the optical density reading along Y-axis. From the graph, the unknown sample concentration was calculated and the results were expressed as mg of rutin equivalents/g dry weight of extract.

2.7. Estimation of Total Antioxidant Capacity (Raghu et al., 2011)

Exactly 0.2 ml of plant leaves extract was added to 2 ml of phosphor molybdenum reagent solution (0.6 M sulfuric acid, 28 mM sodium phosphate and 4mM ammonium molybdate) were incubated at 95°C for 90 minutes. The temperature of the tubes were reduced to room temperature and the absorbance was measured at 695 nm using UV/VIS Spectrophotometer. The antioxidant capacity was expressed as ascorbic acid equivalent.

Total antioxidant activity is calculated by following formula

Ascorbic acid equivalent $(\mu M/g) = (T/S)*C*(V/P)*(RS/E)*(1*MW)$

    T: OD of test solution.

    S: OD of standard.

C: Concentration of test (μg).

V: Volume of solvent used for extraction (ml).

P: Amount of powder (g).

RS: Volume of reagent solution (ml).

E: Volume of extract (ml).

MW: Molecular weight of ascorbic acid (176-13 g/g mol).

2.8. Estimation of Free Radical Scavenging Activity by DPPH Method (Yamaguchi *et al.*, 2008)

DPPH solution (3ml) was added to 1 ml of samples (1-10 mg/ml). The solution in the tubes was vortexed and incubated in the dark for 30 minutes at room temperature and the decrease in the absorbance was measured at 517 nm. The control test tube contains an equal volume of DPPH in methanol instead of extract.

The DPPH radical-scavenging activity in terms of percentage was calculated according to the following equation.

DPPH scavenging activity (per cent) = [(control – sample)/control]*100 per cent

2.9. Estimation of ABTS Cation Radical Scavenging Assay (Sanchez-Moreno, 2002)

ABTS$^+$ assay is based on the inhibition of the absorbance of the radical cation ABTS$^+$, which has a characteristic long wavelength absorption spectrum (Sanchez-Moreno, 2002).

ABTS radical was produced by reacting ABTS solution (7 mM) with ammonium persulfate (2.45 mM).The mixture was allowed to stand in the dark room temperature for 12-16 hours to develop a dark coloured solution.The absorbance of the solution was measured at 745 nm, this stock solution was diluted with methanol to give a final absorbance value of around 0.7 (+ or – 0.02) and equilibriated at 30°C. Different concentrations of the sample (50-250 μg/ml) were prepared by dissolving the extracts in water. About 0.3 ml of the sample was mixed with 3 ml of ABTS working standard in a microcuvette. The decrease in the absorbance was measured at 745 nm after mixing the solution in uniform time interval 3 minutes. A solution of ABTS working standard and 0.3 ml of methanol was used as the control and about 3 ml of methanol was used as blank.

The ABTS cation radical-scavenging activity in terms of percentage was calculated according to the following equation.

ABTS cation radical scavenging activity (per cent) = [(Acontrol – Asample)/Acontrol]*100

2.10. Ferric Ion Reducing Antioxidant Power (FRAP) Assay (Bharathikumar *et al.*, 2008)

Five gram of dried leaves (fresh leaves were air dried in the incubator at 37°C for two days) was weighed and mixed with 50 ml of water in an Erlenmeyer flask and was then incubated overnight at room temperature on an orbital shaker maintained at fixed rpm. The contents were filtered with whatmann filter paper or centrifuged and the filtrate was collected. The solvent in the filtrate was evaporated and 50 mg of the dried powder was dissolved in 50 ml of distilled water. From this, the samples have taken for experimental analysis.

To 2.5 ml of the extract 2.5 ml phosphate buffer (0.2 M, pH 6.6) and 2.5 ml of 1 per cent potassium ferricyanide were added. The resulting mixture was boiled in a water bath at 50°C for 20 minutes, then contents were rapidly cooled, mixed with 2.5 ml of 10 per cent (W/V) trichloro acetic acid. Then the contents were centrifuged at 3000 rpm for 10 minutes. From this, 2.5 ml of supernatant was pipetted out to mixture containing 2.5 ml distilled water and 0.5 ml of 0.1 per cent ferric chloride. The contents in the tube was mixed well and allowed to stand for 10 minutes. The increase in the absorbance at 700 nm was used to measure the reducing power of the plant extract.

2.11. Modified Ferric Ion Reducing/Antioxidant Power Assay (Modified FRAP Assay)

To 0.1 ml of the extract 0.9 ml of 96 per cent ethanol, 5 ml of distilled water, 1.5 ml of 1M HCl, 1.5 ml of 1 per cent potassium ferricyanide, 0.5 ml of 1 per cent SDS and 0.2 per cent ferric chloride was added. The mixture was kept in a water bath maintained at 50°C for 20 minutes. After incubation, tubes were rapidly cooled and mixed well and an increase in the absorbance was read at 750 nm was used to measure the reducing power of the plant extract.

3. Results and Discussion

Some of the green leafy vegetables undertaken in the present investigation are under-exploited, but they are potent sources of natural antioxidants like vitamins, carotenoids, flavonoids and phenols. The selected plant species include four green leafy vegetables *Amaranthus gangetics, Basella alba, Solanum nigrum* and *Sesbania grandiflora* are commonly consumed as green leafy vegetables and believed to have beneficial effects.

3.1. Qualitative Phytochemical Analysis of Plant Extracts

The preliminary phytochemical analysis of plant extracts was carried out and the phytochemicals was present. All the four green leafy vegetables (*Amaranthus gangetics, Basella alba, Solanum nigrum* and *Sesbania grandiflora*) showed health promoting phytochemicals.

3.2. Estimation of Total Phenols

The obtained results indicate the level of total phenols is highest in *Sesbania grandiflora* having total phenol content of 28.18 mg TAE/gDW. This suggests total phenols may be responsible for the free radical scavenging activity of the plant while *Amaranthus gangetics* showed the lowest level of phenolic content of 17.24 mg TAE/g DW. The level of total phenols in *Sesbania grandiflora* has been reported earlier as having 3.01 mg TAE/g DW for ethanol extracts (Shyamalagowri and Vasantha, 2010). The levels of total phenols present in Green leafy vegetables shown in Figure 17.1.

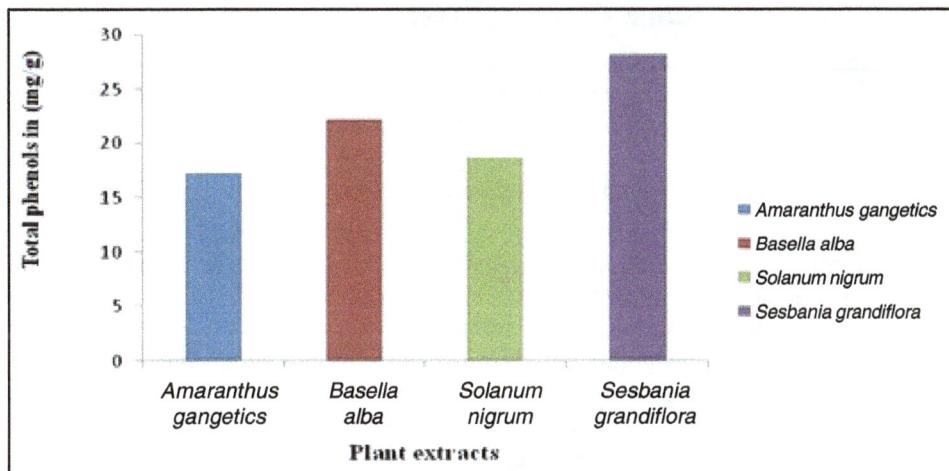

Figure 17.1: Levels of phenolics present in green leafy vegetable.

The antioxidative properties of some vegetables are partly due to the low molecular weight phenolic compounds, which are known to be potent as antioxidants (Huda *et al.*, 2009).

3.3. Estimation of Total Flavonoids

The obtained results indicate the levels of flavonoids are the highest in *Sesbania grandiflora* of 13.54 mg/g. This suggests that flavonoids may be responsible for the free radical scavenging activity of the plant. On the other hand, *Basella alba* showed the lowest levels of flavonoids of 5.7 mg RE/g DW. The results clearly indicated that the *Sesbania grandiflora* is responsible for its antioxidant activity. The amount of total flavonoids in *Basella alba* has been reported as 26.53 mg RE/g DW (Olajire and Azeez2011). The levels of total flavonoids present in the four green leafy vegetables shown in Figure 17.2.

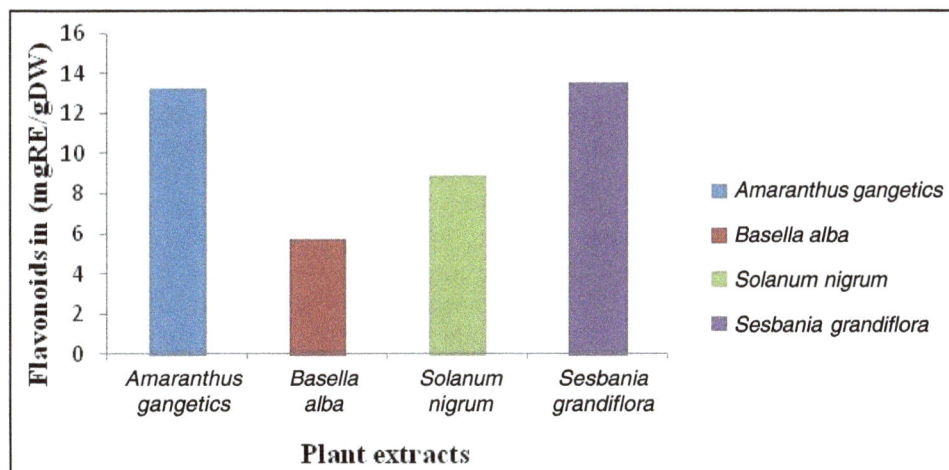

Figure 17.2: Levels of flavonoids present in green leafy vegetables.

3.4. Total Antioxidant Capacity Assay of Ethanol and Aqueous Extracts

The results indicate that the total antioxidant capacity of *Sesbania grandiflora* showed highest total antioxidant capacity (6.009 µM/g). On the other hand, *Basella abla* showed the lowest total antioxidant capacity (1.914 µM/g). The total antioxidant capacity of *Basella alba* has been reported as 4.74 (µM/g).

The results indicate that the total antioxidant capacity of *Amaranthus gangetics* showed highest total antioxidant capacity (4.458 µM/g). On the other hand, *Basella abla* showed the lowest total antioxidant capacity (0.669 µM/g).The total antioxidant assay gives an estimate of the overall antioxidant potential of the plant. There is a formation of phosphomolybdenum complex the intensity of which indicates the potential of the plant as a scavenger of free radicals. The total antioxidant capacity of ethanol extracts of green leafy vegetables shown in Figure 17.3. The total antioxidant capacity of aqueous extracts of green leafy vegetables shown in Figure 17.4.

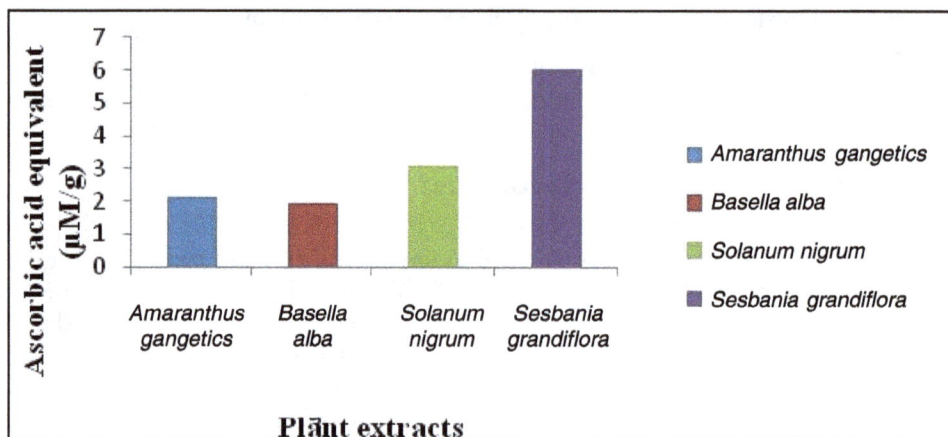

Figure 17.3: Total antioxidant activity of ethanol extracts of green leafy vegetables.

3.5. DPPH Radical Scavenging Activity

The free radicals scavenging activity of DPPH of ethanol extracts of green leafy vegetables shown in Figure 17.5.The values across the concentrations indicate that the ethanol extracts of the plant *Solanum nigrum* shows more potent in neutralizing DPPH free radical. The ethanol extract shows an inhibition of 64-82 per cent in the concentration range of 50-250 µg/ml, while the ethanol extracts of *Basella alba* shows least inhibition of 25-50 per cent in the concentration range of 50-250 µg/ml.The free radicals scavenging activity of *Sesbania grandiflora* has been reported as 18.03 per cent -53.12 per cent for the concentration range (50-250) µg/ml (Shyamala Gowri and Vasantha, 2010).DPPH is a relatively stable free radical. The assay is based on the measurements of the antioxidants' ability to scavenge the stable radical DPPH. The proton radical scavenging action is known as an important mechanism of auto-oxidation. DPPH was used to determine the proton radical scavenging action of the

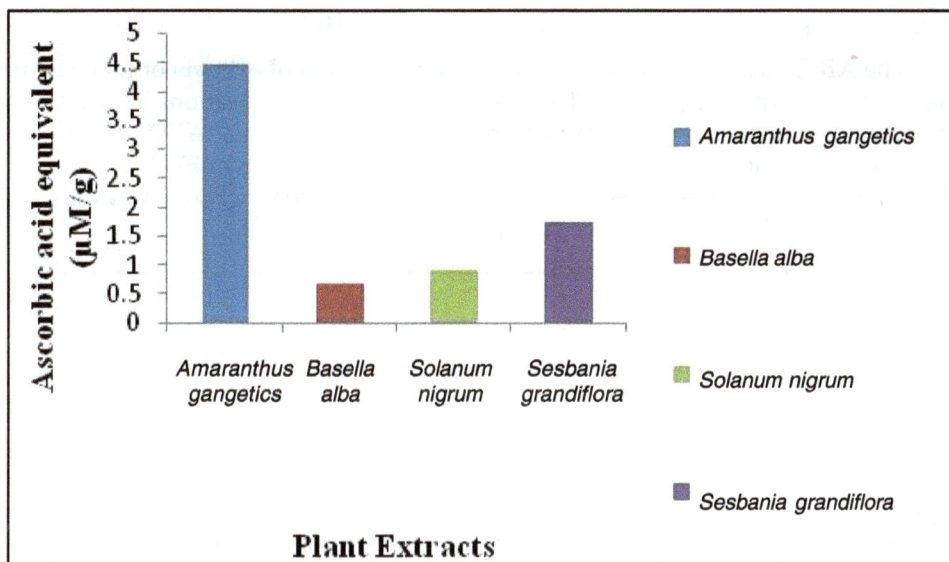

Figure 17.4: Total antioxidant capacity of aqueous extracts of green leafy vegetables.

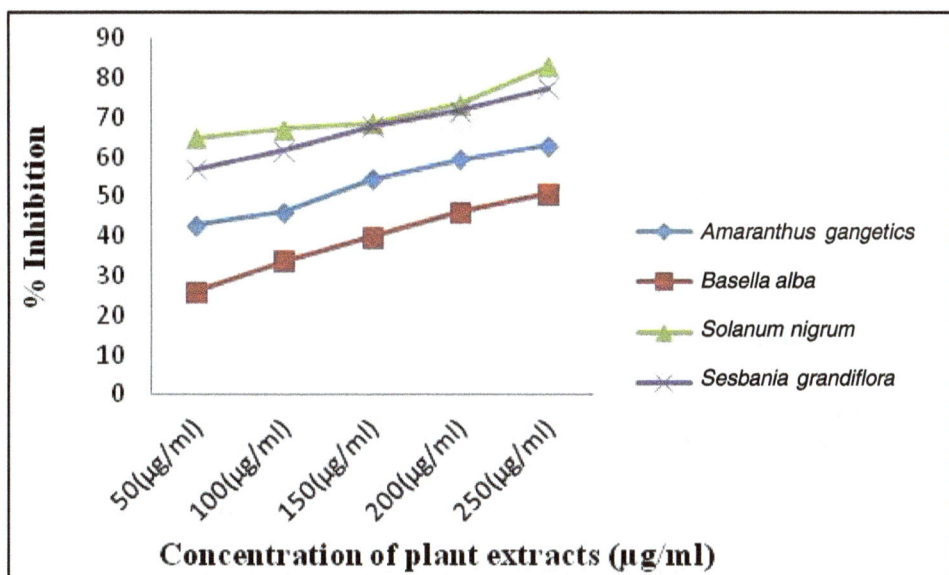

Figure 17.5: Free radicals scavenging activity of DPPH of ethanol extracts of green leafy vegetables.

methanol and acetone extracts of the plants. It shows a characteristic absorbance at 517 nm. The purple color of the DPPH solution fades rapidly when it encounters proton radical scavengers (Yamaguchi *et al.*, 1998).

3.6. ABTS Cation Radical Scavenging Activity

The ABTS cation radical scavenging activity of ethanol extracts of green leafy vegetables shown in Figure 17.6.The values across the concentrations indicate that the ethanol extract of *Basella alba* is more potent in neutralizing ABTS cation free radicals. The ethanol extract of *Basellaalba* shows an inhibition of 37-85 per cent in the concentration range of 50-250 µg/ml, while *Solanum nigrum* shows lowest inhibition of 28-84 in the same concentration range.

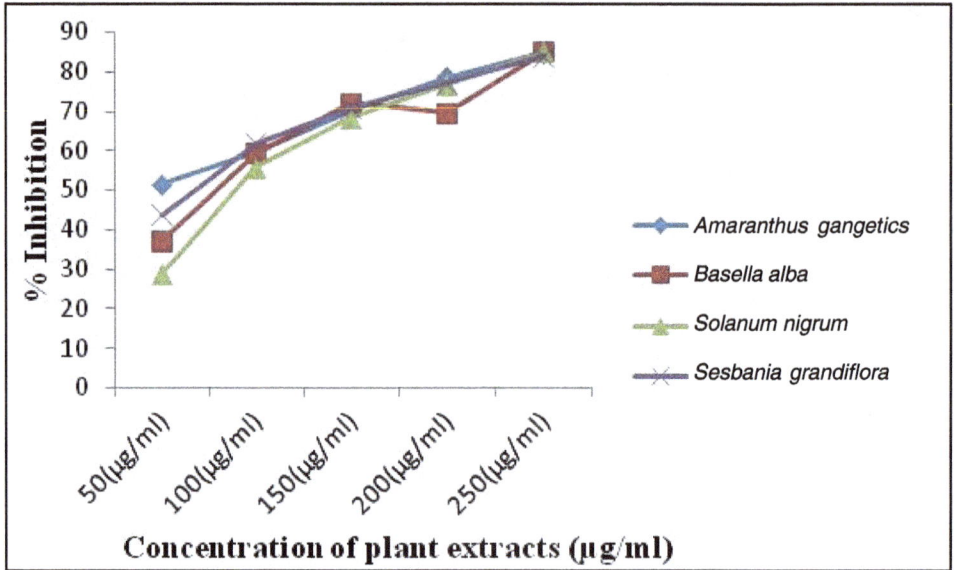

Figure 17.6: ABTS cation radical scavenging activity of ethanol extract of green leafy vegetables.

3.7. Antioxidant Activity of FRAP Assay

The antioxidant activity of FRAP assay by ethanol extracts of green leafy vegetables in the concentration of 100 (µg/ml) are shown in Figure 17.7. Among the four green leafy vegetables *Sesbania grandiflora* showed relatively higher antioxidant activity of absorbance at 700 nm of 2.471, while *Amaranthus gangetics* showed lowest of 0.427 in the concentration of 100 (µg/ml). The Antioxidant activity of FRAP Assay by ethanol extracts of *Sesbania grandiflora* has been reported as having an absorbance at 700 nm of 1.432 (Shyamala Gowri and Vasantha, 2010).

The ferric reducing antioxidant power assay is carried out to determine the ability of the plant extracts to scavenge free radicals by donating electrons. The greater the absorbance, the greater the reducing potential of the plant extract.

3.8. Antioxidant Activity of Modified FRAP Assay

The antioxidant activity of modified FRAP assay by ethanol extracts of green leafy vegetables in the concentration of 100 (µg/ml) shown in Figure 17.8. Among the

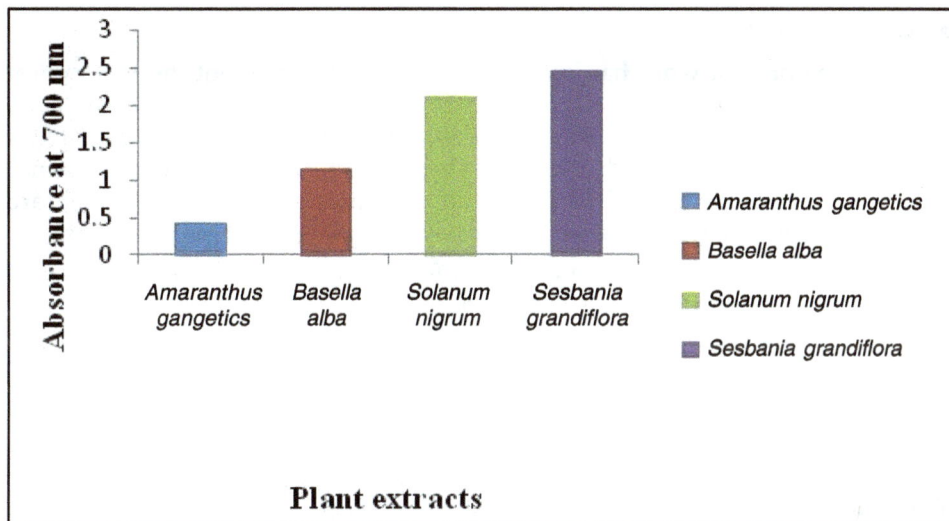

Figure 17.7: Antioxidant activity of FRAP Assay by ethanol extracts of green leafy vegetables.

four green leafy vegetables *Sesbania grandiflora* showed relatively higher antioxidant activity, while *Amaranthus gangetics* showed lowest.

The modified ferric reducing antioxidant power assay is carried out to determine the ability of the plant extracts to scavenge free radicals by donating electrons. The greater the absorbance, the greater the reducing potential of the plant extract.

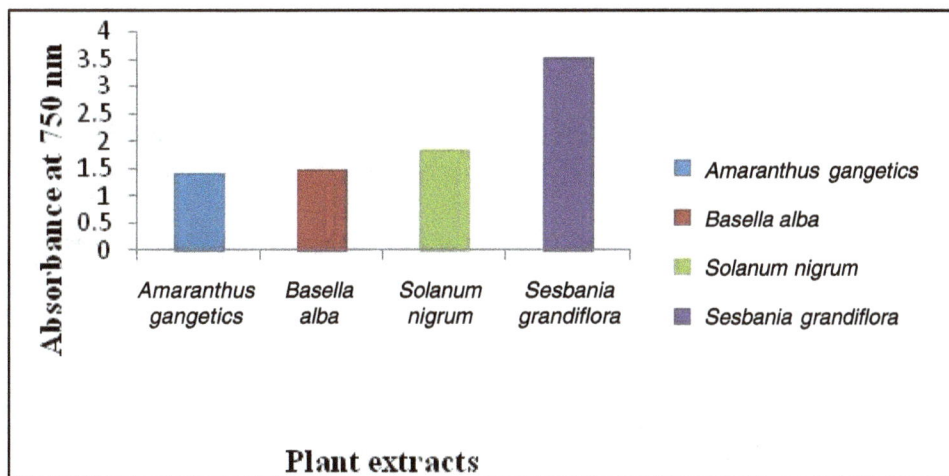

Figure 17.8: Antioxidant activity of modified FRAP assay by ethanol extracts of green leafy vegetables.

5. Conclusion

☆ The present work has been undertaken to study about the free radical scavenging ability of four species of green leafy vegetables, namely, *Amaranthus gangetics, Basella alba, Solanum nigrum* and *Sesbania grandiflora*. The preliminary study carried out in our work has confirmed the presence of phytochemicals like flavonoids, phenols and saponins which are responsible for the free radical scavenging potential of the plant extracts.

☆ The plant species were analyzed for the presence of the various non-enzymatic antioxidants like flavonoids and total phenolics.

☆ Among the selected green leafy vegetables, *Sesbania grandiflora* showed highest potential free radical scavenging activity. It also showed highest phenol and flavonoid contents, which are responsible for the free radical scavenging activity.

References

Bandyopathyay, U., Das, D. and Banerjee, R.K., 1999. Reactive oxygen species: Oxidative damage and pathogenesis. *Current Sci.,* 77: 658.

Bharathikumar, V.M., Satishkumar, T., Shanmugam, S. and Palvanan, T., 2008. Evaluation of antioxidant properties of *Canthium parviflorumlam* Leaves. *Natural Product Radiance,* 7: 122–126.

Chitravadivu, C., Manian, S. and Kalaichelvi, K., 2009. Antimicrobial studies on secleted medicinal plants, Erode region, Tamil Nadu, India. *Middle-East J. of Sci. Research,* 4(3): 147–152.

Demple, B. and Harrison, L., 1994. Repair of oxidation damage to DNA: Enzymology and biology. *Ann. Rev. Biochem.,* 63: 915.

Gella, A. and Durany, N., 2009. Oxidative stress in alzhimer disease. *Cell Adh. Migr.,* 13.

Gupta, S. and Prakash, J., 2009. Studies on Indian green leafy vegetables for their antioxidant activity. *Journal of Plant Foods for Human Nutrition* (Formerly *Qualitas Plantarum*), 64: 39–45.

Halliwell, B. and Gutteridge, J.M.C., 1989. In: *Free Radicals in Biology and Medicine.* Clarendon Press, Publ., Oxford.

Hiner, A., Raven, E., Thorneley, R., García-Cánovas, F. and Rodríguez-López J., 2002. Mechanisms of compound I formation in heme peroxidases *Journal of Inorganic Biochemistry,* 91: 27–34.

Hollman, P.C.H., 2001. Evidence for health effect of plants phenols: Local or systemic effects. *J. Sci. Food. Agric.,* 81: 842–852.

Hayek, M.G., 2000. Dietary vitamin E improves immune function in cats. *Recent Advances in Canine and Feline Nutrition,* 3: 555–564.

Holt, E.M., Steffen, L.M., Moran, A., Basu, S., Steinberger, J., Ross, J.A., Hong, C.P. and Sinaiko, A.R., 2009. Fruit and vegetable consumption and its relation to markers of inflammation and oxidative stress in adolescents. *J. Am. Diet. Assoc.*, 109: 414–421.

Huda, F.N., Noriham, A., Norrakiah, A.S. and Babji, A.S., 2009. Antioxidant activity of plants methanolic extacts containing phenolic compounds. *African J. Biol.*, 8: 484–489.

Miller, N.J., Rice-Evans, C.A. and Paganga, G., 1997. Antioxidant properties of phenolic compounds. *Trends in Plant Science*, 2: 52–159.

Raghu, K.L., Ramesh, C.K., Srinivasa, T.R. and Jamuna, K.S., 2011. Total antioxidant capacity in aqueous extracts of some common vegetables. *Asian J. Exp. Boil. Sci.*, 2: 58–62.

Sanchez-Moreno, C., 2002. Methods used to evaluate the free radical scavenging activity in foods and biological system. *Food Sci. Tech. Int.*, 8: 122.

Sargeant, L.A., Khaw, K.T., Bingham, S., Day, N.E., Luben, R.N., Oakes, S., Welch, A. and Wareham, N.J., 2001. Fruits and vegetable intake and population glycosylated haemoglobinlevels: The EPIC–norfolk study. *Eur. J. Clin. Nutr.*, 55: 342–348.

Scott, R. and Slater, T.F., 1981. Free radical scavenging activity of catechin and other flavonoids. *Recent Advances in Lipid Peroxidation and Tissue Injury*, pp. 233–244.

Shyamala Gowri, S. and Vasantha, K., 2010. Free radical scavenging and antioxidant activity of leaves from Agathi (*Sesbania grandiflora*). *American-Eurasian Journal of Scientific Research*, 5: 114–119.

Singleton, V.L. and Rossi, J.A., 1965. Colorimetry of total phenolics with phosphomolybdic-phosphotungstic acid reagents. *Am. J. Enol. Vitic.*, 16: 144–158.

Yamaguchi, R., Tatsumi, M.A., Kato, K. and Yoshimistu, U., 1958. Effect of salts and fructose on the autooxidation of methyl linoleate in emulsions. *Agr. Biol. Chem.*, 52: 849–850.

Zhishen, J., Mengcheng, T. and Jianming, W., 1999. *Food Chemistry*, 64(4): 555–559(5).

Scientific Basis of Herbal Medicine (2013)
Editor: Dr. Parimelazhagan Thangaraj
Published by: DAYA PUBLISHING HOUSE, NEW DELHI

Pages 163–171

Chapter 18

Medicinal Plant Diversity of the Sacred Groves in Mahe, U.T. of Puducherry

*K. Sasikala[1], G. Pradeepkumar[1], C.C. Harilal[2],
E. Girishkumar[1] and C.P. Ravindran[1]*

[1]*P.G. Department of Plant Science, Mahatma Gandhi Govt. Arts College,
Mahe – 673 310, U.T. of Puducherry, India*
[2]*University of Calicut,
Calicut University, Calicut – 673 635, Kerala, India*

1. Introduction

Medicinal plants play a significant role in the day-to-day life of human beings. Majority of them have become rare and threatened due to over-exploitation, habitat destruction and other developmental activities. Habitat destruction is considered as one of the severe threats to biodiversity. The natural habitats of many of the native species of flora and fauna are under tremendous pressure and declining at a rapid rate. They are now confined only to protected areas such as national parks, wildlife sanctuaries, reserve forests, sacred groves, and various wildlife reserves in India and across the world. Sacred groves are small patches of native vegetation traditionally protected by local communities. They act as repositories of various floral and faunal life-forms.

Many of the sacred groves are abode of a number of rare medicinal plants. The sacred groves play a pivotal role in the conservation of biodiversity in general and medicinal plants in particular. This is due to the belief system and taboos associated with the groves and restriction to the local inhabitants to enter into the groves. A review of literature revealed that a comprehensive documentation of sacred groves of Mahe is lacking hitherto. Therefore the present study was carried out to document the sacred groves of Mahe and their biodiversity with special reference to diversity of medicinal plants. It is also aimed to generate a primary database on the medicinal plant wealth of sacred groves of Mahe.

2. Study Area

Mahe is one of the four administrative districts of the Union Territory of Puducherry. It lies on the west coast of Indian Peninsula between Kozhikode and Kannur districts of Kerala and is located between 1142' – 1143' N and 7531' – 7533' E and occupies an area of 9 km². It also forms a part of biodiversity-rich Western Ghats. Mahe consists of three regions namely, Mahe proper, Kallayi and Naluthara enclaves. Mahe proper, a small town lies on the river Mahe. The Naluthara enclave lies between the Ponniyam River in the north and Kozhikode-Thalassery road in the south and Kallayi is situated between these areas.

3. Methodology

Regular field surveys were undertaken for a period of two years (2010–12). During the field surveys assessment of the floral diversity of the groves was carried out. The representative plant species were collected, processed, identified and documented. Plants that have been collected during the present study are identified using floras (Gamble, 1915–36; Ramachandran and Nair, 1988; Sasidharan, 2004) and other relevant literature. The herbarium specimens were prepared following standard herbarium techniques (Fosberg and Sachet, 1965) and are deposited at herbarium of Mahatma Gandhi Government Arts College, Mahe, for reference. Details pertaining to the medicinal uses of plants were gathered by interviewing the stakeholders of the sacred groves, the priests and elderly persons inhabiting in the vicinity of the sacred groves. Botanical name of the medicinal plant species with authority, family, habit, local name, parts used, mode of administration and diseases treated are provided.

4. Results

In the present study a total of 19 sacred groves have been located in Mahe. These sacred groves harbour various life-forms of angiosperms such as herbs, shrubs, trees, climbers, epiphytes and parasites, besides few species of pteridophytes, bryophytes, gymnosperms, macro fungi and lichens. These sacred groves also possess endemic and threatened plant species and play a vital role in conserving the biodiversity. A total of 324 angiosperm taxa belonging to 95 families have been recorded from these groves. Of these 59 species belonging to 58 genera and 35 families are used by the local inhabitants for curing various ailments.

4.1. Enumeration

1. *Achyranthes aspera* L. (Amaranthaceae): Herb. *Kadaladi*

Uses: The decoction of the whole plant is used to control swellings and mixed along with honey and *Plectranthus amboinicus* (*Panikoorka*) to treat respiratory disorders. It is also used to control ear and stomach related problems.

2. *Aegle marmelos* (L.) Corrêa (Rutaceae): Tree. *Koovalam*

Uses: Root paste is used for gastric problems. Three leaves of *Aegle marmelos* along with the leaves of *Mimosa pudica* and *Murraya koenigii* are chewed in empty stomach to cure diabetes. Leaves act as an insecticide. Plant is also used for epilepsy.

3. *Alstonia scholaris* (L.) R. Br. (Apocynaceae): Tree. *Ezhilampala*

Uses: Latex is used to loosen the bowels and is applied externally to cure wounds. The decoction made from bark is used to control fever and skin diseases.

4. *Amorphophallus campanulatus* Decne. (Araceae): Herb. *Chena*

Use: Corm is consumed regularly to control swelling in piles.

5. *Anacardium occidentale* L. (Anacardiaceae): Tree. *Kasumavu*

Uses: The paste of fruit and bark is applied externally to cure arthritis. Fruit juice is taken to cure stomach disorders.

6. *Andrographis paniculata* (Burm.f.) Wall. ex Nees (Acanthaceae): Herb. *Kiriyathu*

Uses: The juice of whole plant is used against inflammation of liver and mixed along with *Justicia adhatoda* to treat fever and wheezing.

7. *Annona squamosa* L. (Annonaceae): Tree. *Aathachakka*

Uses: The decoction of the leaf is administered orally to control fissures in children. The pulp of the ripened fruit is applied externally on wounds to break open the pussy ulcers for easy curing.

8. *Asparagus racemosus* Willd. (Asparagaceae): Herb. *Sathavari*

Uses: Tuber paste along with milk is given for treating diarrhoea. Tuber is pickled and consumed.

9. *Azadirachta indica* A. Juss. (Meliaceae): Tree. *Ariyaveppu*

Uses: Leaves along with turmeric powder is boiled in water and taken bath to cure skin diseases. It acts as a cooling agent also.

10. *Bacopa monnieri* (L.) Wettst. (Scrophulariaceae): Herb. *Neer Brahmi*

Uses: Whole plant juice mixed with milk is consumed to cure menstrual problems and stomach disorders. It is also used as brain tonic to enhance memory power.

11. *Biophytum sensitivum* (L.) DC. (Oxalidaceae): Herb. *Mukkuti*

Uses: Leaf paste is applied to cure wounds. Whole plant is ground and the paste along with cold water or milk is taken orally for a week to cure excess bleeding and strengthen the uterus.

12. *Boerhavia diffusa* L. (Nyctaginaceae): Herb. *Thazhuthama*

Uses: Whole plant is used as vegetable, against iron deficiency, urinary tract infection, arthritis and acts as a rejuvenator.

13. *Bombax ceiba* L. (Malvaceae): Tree. *Ilavu*

Uses: Leaves are crushed and applied externally to cure wounds.

14. *Borassus flabellifer* L. (Arecaceae): Tree. *Karimpana*

Uses: Tender leaves are pounded and given orally against ortho diseases. Roots are used for nerve rejuvenation

15. *Butea monosperma* (Lam.) Taub. (Fabaceae): Tree. *Chamatha*

Uses: The seeds are made in to a paste and mixed with butter milk and consumed to control pinworm infection. The decoction from bark is used to control excessive bleeding.

16. *Cardiospermum halicacabum* L. (Sapindaceae): Herb. *Uzhinja*

Uses: The leaves are fried and applied on stomach for menstrual disorders. Leaf paste is used as thali.

17. *Carica papaya* L. (Caricaceae): Tree. *Papaya*

Uses: The fruit when consumed helps in promoting digestion, prevents anemia and has anti-inflammatory properties. Fruit skin paste is used to cure old skin lesions and itchy dermatitis.

18. *Chromolaena odorata* (L.) R.M. King and H. Rob. (Asteraceae): Shrub. *Communist pacha*

Uses: The leaf paste is used against cut and injury. The extract of tender twig is boiled in water and taken bath or applied externally to relieve joint pain that occurs after chikungunya disease.

19. *Centella asiatica* (L.) Urb. (Apiaceae): Herb. *Muthil*

Uses: The plant is used to enhance the memory power, taken along with butter to cure epilepsy in children.

20. *Citrus limon* (L.) Burm.f. (Rutaceae): Tree. *Illumbichinarekam, Cherunarekam*

Uses: Juice of leaves is mixed in water and taken bath to get rid-off cold. It also acts as a blood purifier and enhances digestion. The fruit juice mixed with salt is taken in empty stomach in the morning to cure stomach pain.

21. *Clitoria ternatea* L. (Fabaceae): Herb. *Sankupushpam*

Uses: The root juice along with coconut milk is used as laxative. The root along with powder of asafoetida, *Zingiber officinale*, pepper and rhizome is ground and mixed with milk and given orally at night for 10 days to expel toxic substances from the body.

22. *Curculigo orchioides* Gaertn. (Hypoxidaceae): Herb. *Nilappana*

Uses: Powdered rhizome mixed with honey is used against urinary infection. Raw tuber is ground and mixed with milk is used against jaundice.

23. *Curcuma longa* L. (Zingiberaceae): Herb. *Manjal*

Uses: Pounded onion along with turmeric powder is applied to cure insect bite. Turmeric powder mixed with water is taken orally for reducing diabetes.

24. *Desmodium triflorum* (L.) DC. (Fabaceae): Herb. *Nilamparanda*

Uses: The juice of whole plant along with coconut milk is boiled and cooled and is applied on face to ward off pimples.

25. *Eclipta prostrata* (L.) L. (Asteraceae): Herb. *Kanjunni*

Uses: Plant juice is used as appetizer, for night blindness and in the preparation of hair oil. The leaves along with Hibiscus flowers and leaves of *Lawsonia inermis* are made into a paste and added to coconut oil and applied externally for hair fall.

26. *Elephantopus scaber* L. (Asteraceae): Herb. *Anachuvadi*

Uses: The whole plant decoction is used against dysentery and diarrhea. The powdered rhizome along with unpolished rice mixed with palm jaggery is used to control bleeding piles.

27. *Emilia sonchifolia* (L.) DC. (Asteraceae): Herb. *Muyal chevian*

Uses: The paste of the leaves along with coconut milk, pepper, betel leaf is mixed and applied on head to remove tonsillitis. Leaves are consumed in empty stomach to cure diarrhoea and used in the preparation of hair oil.

28. *Erythrina variegata* L. (Fabaceae): Tree. *Murikku*

Uses: Tender leaf juice is applied on the head as well as inhaled to cure bleeding of nose.

29. *Globba sessiliflora* Sims (Zingiberaceae): Herb. *Kolachanna*

Uses: Rhizome is carminative, expectorant, laxative and reduces flatulence.

30. *Gloriosa superba* L. (Colchicaceae): Herb. *Malattamara*

Uses: The rhizome paste is applied externally for insect bite. The paste of leaf is used against hair fall due to fungal infection. The rhizome paste is used for smooth delivery during child birth.

31. *Helicteres isora* L. (Malvaceae): Tree. *Edampiri Valampiri*

Uses: The decoction of the root is used to cure diabetes, stomachache and pinworms. The fruits are soaked in oil and applied as drops internally to cure ear problems.

32. *Hibiscus rosa-sinensis* L. (Malvaceae): Shrub. *Chembarathi*

Uses: Three flower buds of Hibiscus ground and mixed with rice water are taken orally for lowering blood pressure. Root extract is consumed against urinary tract infection. Flower juice is used for sunburns.

33. *Hydnocarpus pentandrus* (Buch.-Ham.) Oken (Achariaceae): Tree. *Marotti*

Uses: Oil obtained from the seed is used to cure leprosy, sprain, against skin diseases. Fruit pulp is used to cure eye disease.

34. *Indigofera tinctoria* L. (Fabaceae): Shrub. *Amari*

Uses: Leaf juice along with honey is used to control jaundice. Plant paste is used externally against scorpion bite.

35. *Justicia adhatoda* L. (Acanthaceae): Shrub. *Aadalodakam*

Uses: Leaf juice along with cow's milk is consumed for cough and asthma.

36. *Leucas aspera* (Willd.) Link (Lamiaceae): Herb. *Thumba*

Uses: Leaf juice is taken orally to stop hiccups and enhances memory power. Flowers boiled in milk are given to children for stomach disorders.

37. *Maranta arundinacea* L. (Marantaceae): Herb. *Koova*

Uses: The powdered rhizome is used to cure dysentery.

38. *Microstachys chamaelea* (L.) Müll.Arg. (Euphorbiaceae): Shrub. *Kodiyavanakku*

Uses: Plant extract along with coconut milk and *Cuminum cyminum* seeds is boiled and the residue is removed and is taken orally for rheumatism and arthritis.

39. *Mimosa pudica* L. (Fabaceae): Herb. *Thottavadi*

Uses: Leaf paste is applied externally to cure wounds, cuts and headache. Root is grounded into paste along with goat's milk is used against respiratory diseases and acts as blood purifier.

40. *Moringa oleifera* Lam. (Moringaceae): Tree. *Muringa*

Uses: The leaves are washed with turmeric powder and boiled and made into a pudding and consumed for lowering blood pressure. The leaves, flowers and fruits are cooked and consumed for treatment against constipation, urinary tract infection and leukemia.

41. *Murraya koenigii* (L.) Spreng. (Rutaceae): Tree. *Karivepila*

Uses: Leaves along with pepper powder and buttermilk are taken orally in the morning and evening for curing mouth ulcers. Leaves as a source of Vitamin A are dried and pounded along with *Cajanus cajan*, *Phaseolus mungo*, chilli and salt and consumed as a side dish for breakfast. Leaf paste mixed in buttermilk is taken orally to cure piles. The hair oil prepared from the leaves is used to cure dandruff and small boils of head.

42. *Musa paradisiaca* L. (Musaceae): Herb. *Nenthravazha*

Uses: Fruits are dried and powdered and mixed with milk and given as a nutrient food to infants and the old. Leaf sheath is pounded and tied over the burns for a week to cure the associated symptoms. Stem juice is consumed against excess bleeding. Tubers are used for tooth disease.

43. *Myristica fragrans* Houtt. (Myristicaceae): Tree. *Jathika*

Uses: An amount of 125 gm of *Myristica* boiled and ground in one litre of coconut oil and applied for arthritis and wounds.

44. *Ocimum tenuiflorum* L. (Lamiaceae): Undershrub. *Tulsi*

Uses: Leaf extract is applied externally for skin irritation. It is also consumed along with black pepper, ginger and honey to cure cough and cold.

45. *Oxalis corniculata* L. (Oxalidaceae): Herb. *Puliyaaral*

Uses: About 10 gm of whole plant along with one glass of butter milk is boiled and taken to control bleeding. Leaf juice is mixed with onion to cure warts.

46. *Phyllanthus amarus* L. (Phyllanthaceae): Herb. *Keezharnelli*

Uses: Roots and leaves along with seeds of *Cuminum cyminum* are made into a paste and a small ball size of the same dipped in cow's milk is taken early morning in empty stomach to cure jaundice and urinary infections. Plant juice is used as a shampoo for a month to reduce dandruff and hair falling.

47. *Phyllanthus emblica* L. (Phyllanthaceae): Tree. *Nellika*

Uses: Fruits are antioxidant and anti-ageing properties and are pickled and consumed. They are effective in controlling dehydration. Fruit ground along with jaggery is consumed to increase body resistance.

48. *Piper nigrum* L. (Piperaceae): Herb. *Kurumulaku*

Uses: Pepper powder along with Tulsi leaf is boiled and given for curing cough. Pepper paste along with milk applied on forehead to remove blood clotting and for active circulation of blood.

49. *Portulaca oleracea* L. (Portulacaceae): Herb. *Kozhuppa*

Uses: Whole plant is used to maintain normal body temperature. It is also used in constipation and urinary infection.

50. *Psidium guajava* L. (Myrtaceae): Tree. *Perakkai*

Uses: Leaves are boiled in water and taken early morning to reduce the sugar level.

51. *Pterocarpus santalinus* L.f. (Fabaceae): Tree. *Rakthachandanam*

Uses: Bark paste is applied externally to cure headache, ulcers and pimples. A mixture of powdered red sandal wood along with jaggery and rice water is used to control intestinal ulcers.

52. *Saraca asoca* (Roxb.) de Wilde (Fabaceae): Tree. *Ashokam*

Uses: Bark and flowers are used against cuts, menstrual problems and uterus dysfunction. It also gives strength to bones and used in the treatment of skin ailments.

53. *Scoparia dulcis* L. (Scrophulariaceae): Herb. *Kallurki*

Uses: A gooseberry-sized plant paste is consumed to remove kidney stones and bladder stones.

54. *Sida acuta* Burm.f. (Malvaceae): Herb. *Kurunthotti*

Uses: Root made into kashayam and taken twice a day for secretion of synovial fluid in bones. Root paste is mixed along with milk and oil is applied externally for arthritis.

55. *Spondias pinnata* (L.f.) Kurz (Anacardiaceae): Tree. *Ambazham*

Uses: Fruit are pickled and consumed. Bark paste is used in the treatment of dysentery.

56. *Strychnos nux-vomica* L. (Loganiaceae): Tree. *Kanjiram*

Uses: About 500 mg of seed powder mixed in water is consumed for three days to cure rheumatic pain.

57. *Tabernaemontana divaricata* (L.) R. Br. ex Roem. and Schult. (Apocynaceae): Shrub. *Nadyarvattam*

Uses: Flower extract is applied to cure eye disease and applied externally on wounds.

58. *Tinospora cordifolia* (Willd.) Hook.f. and Thomson (Menispermaceae): Climber. *Chittamruthu*

Uses: About 15 ml of stem juice is taken in the morning and evening for kidney related diseases. Stem paste is applied externally for swellings. Plant also has blood purification properties.

59. *Tragia involucrata* L. (Euphorbiaceae): Herb. *Kodithuva*

Uses: Whole plant along with *Piper longum*, kismis and honey is consumed for respiratory disorders and are boiled in water and taken early morning to reduce the sugar level.

5. Conclusion

The sacred groves are potential conservative areas of flora and fauna. They preserve a good number of native species of flora, especially medicinal plants. The present study also endorses this view. Medicinal plants are invaluable resources of many life saving new drugs. Overexploitation, invasion of exotic species and habitat destruction are posing serious threats to medicinal plants. Preservation of sacred groves is one of the best methods to conserve the native plant species that have been utilized by the native people for their various medicinal properties.

Acknowledgements

The authors wish to express their sincere gratitude to Department of Science, Technology and Environment, Govt. of Puducherry, for financial assistance. They are thankful to the Principal, MGGA College, Mahe, for encouragement and Dr. W. Arisdason, Scientist, Botanical Survey of India, Kolkata, for identification of plants and providing literature. The timely help rendered by department colleagues is also gratefully acknowledged. The authors are also grateful to the stakeholders of sacred groves and the residents of Mahe, for providing information on sacred groves and medicinal properties of plants.

References

Fosberg, F.R. and Sachet, H., 1965. Manual of tropical herbaria. *Regnum Veg.*, Vol. 39. The Netherlands.

Gamble, J.S., 1915–1936. *Flora of the Presidency of Madras.* 11 Parts. (Parts 1–7 by Gamble and 8–11 by C.E.C. Fischer), London. Repr. ed., 1957. Botanical Survey of India, Calcutta.

Ramachandran, V.S. and Nair, V.J., 1988. *Flora of Cannanore.* Botanical Survey of India, Calcutta.

Sasidharan, N., 2004. *Biodiversity Documentation for Kerala. Part 6: Flowering Plants.* Kerala Forest Research Institute, Peechi.

Scientific Basis of Herbal Medicine (2013)
Editor: Dr. Parimelazhagan Thangaraj
Published by: DAYA PUBLISHING HOUSE, NEW DELHI

Pages 173–176

Chapter 19

Insecticidal and Repellent Activities of Medicinal Plant Extracts on Diamond Back Moth, *Plutella xylostella* (Linn.) (Plutellidae : Lepidoptera)

Mariappan Suganthy, Sentrayaperumal Sundareswaran, Periyaswamy Sakthivel and Lakshmanan Nalina

*Tamil Nadu Agricultural University,
Coimbatore – 641 003, Tamil Nadu, India*

1. Introduction

Although self-sufficiency in food grain production has been achieved through green revolution there is an apprehension over the possible paucity of food grains in the near future due to the alarming population growth. This problem is further intensified by the loss of food grains due to insect pests. Indiscriminate use of chemicals has resulted in pesticide resistance, resurgence of target organism or emergence of secondary pests because of destruction of parasitoids and predators, impact on non-target organisms, including humans, environmental pollution through accumulation of pesticides in soil, water and air and residues in agricultural and animal products. Increasing awareness about the deleterious effects of insecticides paved the way for integrated and eco-friendly pest management. One such method is the use of botanical pesticides, which are safe, eco-friendly and can overcome many problems associated with chemical insecticides. As early as 1920, the occurrence of *Plutella xylostella* (Linn.) was reported on different cruciferous crops all over India. Later, its occurrence as a severe pest of crucifers was reported by several workers. This pest has developed resistance to invariably all groups of insecticides including fenvalerate. Hence, botanical pesticides have been used to tackle this pest problem to some extent.

2. Materials and Methods

2.1. Mass Culturing of *Plutella xylostella*

Mass culturing of *Plutella xylostella* was carried out using the mustard seedlings (for oviposition) and cauliflower leaves, at 25°C and 75±5 per cent RH.

2.2. Extraction of Plant Material

The leaves of *Mentha arvensis, Pogostemon patcholi, Tinospora cordifolia, Tylophora asthmatica, Ocimum gratissimum* and *Ruta graveolens* were collected. Ten per cent aqueous extracts of these leaves were prepared by taking 100 grams of these leaf material, making them into paste and the final volume made into 1 litre using distilled water. The extracts were kept as such over night and filtered after 24 hours using filter paper. The extracts prepared were labeled and stored in the refrigerator. Apart from plant extracts, three TNAU neem formulations *viz.*, TNAU NO (Neem Oil) 60 EC (A), TNAU NO (Neem Oil) 60 EC (C) and TNAU NOPO (Neem Oil + Pungam Oil) 60 EC (5 per cent) were also used.

2.3. Bio-assay

For the bio-efficacy test, the extracts were taken in the hand sprayer and sprayed on the potted cauliflower seedlings of 60 days old and each pot was considered as a treatment. To each plant ten *P. xylostella* early third instar larvae were released and covered using mylar film cage. There were ten treatments and each treatment was replicated thrice. The larvae were observed 6, 12, 24, 48, 60 and 72 hours after release (HAR) for mortality. Percentage pupation was worked out and pupae were allowed for adult emergence and larval, pupal and adult malformations were recorded. Number of eggs laid per treated plant was recorded. Hatchability of eggs on different treatments was also recorded.

2.4. Statistical Analysis

The data recorded in this experiment were analysed statistically using completely randomized design. The percentage data were transformed into corresponding angles (arcsin) for statistical interpretation. Where ever there was zero value, data were transformed into square root for statistical interpretation. Duncan's Multiple Range Test (DMRT) was applied to analyse the data. In the table, in each coloumn, means followed by a common letter were not significantly different at 5 per cent level.

3. Results and Discussion

The results of bio efficacy test of medicinal plant extracts on the life stages of *P. xylostella* revealed that at 6, 12 and 24 hours after treatment (HAT) the larval mortality was observed to be nil. At 48 HAT, only TNAU neem formulations recorded maximum mortality of 33.33 and 26.667 per cent. At 60 HAT, TNAU NO 60 EC (A), TNAU NO 60 EC (C) and TNAU NO PO 60 EC were found to be superior with maximum mortality, followed by 20.00 per cent in *Ruta graveolens* which was found to be on par with *Mentha arvensis* and *Tinospora cordifolia*. While at 72 HAT, the treatments were found to be minimum in TNAU NO 60 EC (A), TNAU NO 60 EC (C) and TNAU NOPO 60 EC (C). *Tinospora cordifolia* and *Tylophora asthmatica* recorded

Table 19.1: Bioefficacy of medicinal plant extracts on the life cycle of *Plutella xylostella*.

Treatments		Larval mortality (per cent)			Pupation (per cent)	Pupal or Adult Malformation (per cent)	No. of Eggs/ Treatment (per cent)	Hatchability (per cent)
		48 HAT	60 HAT	72 HAT				
T₁	*Mentha arvensis*	6.66ab (8.85)	13.33ab (17.71)	0.03a (0.19)	80.00b (63.43)	0.03a (0.19)	27.67b (5.36)	99.97a (89.89)
T₂	*Pogostemon patcholi*	13.33ab (17.70)	6.66a (8.85)	6.66a (8.85)	73.33b (59.23)	0.03a (0.19)	27.33b (5.27)	99.97a (89.89)
T₃	*Tinospora cordifolia*	13.33ab (17.71)	13.33ab (17.71)	6.66a (8.85)	66.67bc (54.99)	0.03a (0.19)	26.00b (5.14)	99.97a (89.89)
T₄	*Tylophora asthmatica*	0.03a (0.19)	6.66a (8.85)	0.03a (0.19)	93.33c (81.15)	0.03a0.19	26.00b (5.14)	99.97a (89.89)
T₅	*Ocimum gratissimum*	0.03a (0.19)	0.03a (0.19)	0.03a (0.19)	99.97c (89.89)	0.03a0.19	26.33b (5.17)	99.97a (89.89)
T₆	*Ruta graveolens*	6.66a (8.85)	20.00ab (21.92)	6.66a (8.85)	66.67b (54.99)	0.03a (0.19)	27.33b (5.27)	99.97a (89.89)
T₇	TNAU–NO 60 EC [A]	33.33b (35.09)	40.00b (38.85)	0.03a (0.19)	26.67a (30.78)	0.03a (0.19)	22.33a (4.77)	99.97 (89.89)
T₈	TNAU–NO 60 EC [C]	33.33b (35.09)	33.33b (35.09)	0.03a (0.19)	33.33a (35.09)	0.03a (0.19)	21.00a (4.63)	99.97a (89.89)
T₉	TNAU–NO PO 60 EC	26.67b (30.77)	40.00b (38.85)	0.03a (0.19)	33.33a (35.09)	0.03a (0.19)	21.00a (4.63)	99.97a (89.89)
T₁₀	Untreated control	0.03a (0.19)	6.66a (8.85)	0.03a (0.19)	93.33c (81.15)	0.03a (0.19)	29.00a (5.43)	99.97a (89.89)

* The values are mean of three replications.

* Figures in parentheses (except No. of eggs) are arcsin transformed.

* Figures in parentheses (No. of eggs) are square root transformed.

maximum percentage pupation which was found to be on par with untreated control. Larval-pupal or pupal-adult malformation was observed to be nil and all the treatments were found to be non-significant. TNAU NO 60 EC (A), TNAU 60 EC (C) and TNAU NOPO 60 EC were found to be significantly superior in having oviposition deterrent property and all other treatments were found to be on par with control. With regard to hatchability, the treatments were found to have no significant difference (Table 19.1).

The results of the current study were in accordance with the findings of Ohno and Hirota (1993), Hermawan *et al.* (1994), Deka *et al.* (1998) and Oudia (2000). Ohno and Hirota (1993) controlled *Pratylenchus penetrans* using some antagonistic plants. Antifeedant and antioviposition activities of the fractions of extract from *Andrographis paniculata*, against *Plutella xylostella* was recorded by Hermawan *et al.* (1994). Deka *et al.* (1998) reported that *Clerodendron inerme* and *Polygonum orientale* were found to have antifeedant and repellent properties against tea mosquito bug, *Helopeltis theivora*. Oudia (2000) evaluated some botanicals against *Aspidomorpha miliaris* and *Zonabris pustulata*.

References

Deka, M.K., Singh, Karna and Handique, R., 1998. Antifeedant and repellent properties of *Clerodendron inerme* and Polygonum orientale against tea mosquito bug, *Helopeltis theivora* Waterhouse. *Two and a Bud*, 45(2): 8–10.

Hermawan, W., Kajiyama, S., Tsukuda, R., Fujisaki, K., Kobayashi, A. and Nakasuji, F., 1994. Antifeedant and antioviposition activities of the fractions of extract from a tropical plant, *Andrographis paniculata* (Acanthaceae), against the diamondback moth, *Plutella xylostella* (Lepidoptera: Yponomeutidae). *Applied Entomology and Zoology*, 29(4): 533–538.

Ohno, T.and Hirota, K., 1993. Control of *Pratylenchus penetrans* using some antagonistic plants. *Research Bulletin of the Aichi-ken Agricultural Research Centre* 25, 221–228.

Oudhia, P., 2000. Effects of leaf extracts on metallic coloured tortoise beetle *Aspidomorpha miliaris* F. *Insect Environment*, 5(4): 165.

Oudhia, P., 2000. Evaluation of some botanicals against orange banded blister beetle (*Zonabris pustulata* Thunb.). *Crop Research (Hisar)*, 20(3): 558–559.

Scientific Basis of Herbal Medicine (2013)   *Pages 177–189*
Editor: Dr. Parimelazhagan Thangaraj
Published by: DAYA PUBLISHING HOUSE, NEW DELHI

Chapter 20

Quorum Quenching Activity of *Plumbago zeylanica* Linn.

S. Ashokraj, D. Ramathilagam and V. Brindha Priyadarisini

Department of Microbial Biotechnology,
Bharathiar University, Coimbatore – 641 046, T.N., India

1. Introduction

Plant derived medicinal compounds are alternative source for control of microorganisms. These compounds are safe and have a long history of use in traditional medicine. Detection of mechanism behind this activity throws a light on the use of these compounds. One among the activities reported is quorum quenching.

Bacteria possess global regulatory systems that adapt virulence gene expression to changing environmental conditions during infection (Kong *et al.*, 2006). Among these global regulatory systems, cell-cell communication, also known as quorum sensing (QS), have come to forefront over recent years. QS, or cell-cell communication, describes the regulation of gene expression in response to increasing cell density and thereby enabling the bacteria to adapt to changing environmental conditions, such as a change in nutrient supply, altered oxygen levels and the switch from planktonic to biofilm growth (Otto, 2004). QS plays an essential role in synchronising gene expression and functional co-ordination among bacterial communities (Dong and Zhang, 2005) and is crucial in establishing a well-ordered surface community. Specifically, QS systems play a central role in staphylococcal pathogenesis and appear to influence biofilm development at many of the distinct stages of biofilm formation. QS Systems use small signalling molecules known as autoinducers (AIs). Once the AIs accumulate to a certain threshold level, activation of the QS system occurs and triggers the direct/indirect transcription of target genes (Xu *et al.*, 2006), often including a series of virulence factors. Two QS systems have been identified and characterised in staphylococci, the luxS QS system and the accessory gene regulator (agr) system, both of which regulate several of the biofilm-associated factors of *Staphylococcus aureus* and *Staphylococcus epidermidis* at various stages of biofilm formation.

Costerton *et al.* (1987) defined a biofilm as the accumulation of microorganisms and their extracellular products to form a highly structured bacterial community on a surface. Biofilm formation of *S. epidermidis*, as with other bacteria, is a multistep process and occurs in four distinct phases-attachment (adhesion), accumulation, maturation and detachment. The behaviour of *S.epidermidis* during biofilm development and within the sessile community is greatly influence by cell-to-cell communication, a process known as quorum sensing (QS) which appears to influence biofilm formation at each of the stages of biofilm formation (Kong *et al.*, 2006).

Many plants have been reported to demonstrate Quorum quenching. There are a number of ways to interrupt the QS system. Thus (anti-QS) compounds can be of great interest in the treatment of bacterial infections. A number of quorum-quenching enzymes that hydrolyze AHLs have been identified in bacteria (Dong and Zhang, 2005). To date, the only known anti-QS compounds of non-bacterial origin are halogenated furanones from the red alga *Delisea pulchra* (Manefield *et al.*, 1999). Quorum quenching activity has also been shown in a number of southern Florida seaweeds (Cumberbatch, 2002) and a few terrestrial plants (Teplitski *et al.*, 2000; Gao *et al.*, 2003). However, so far, only a handful of higher plants have been studied and a few reports are available to anti-QS activity in medicinal plants. The plant kingdom has long been a source of medicines and continues to contribute significantly to the development of today's pharmaceuticals (Cragg *et al.*, 1997). The emergence of antibiotic resistance begs the need for novel therapeutics. It has been suggested that targeting the QS system, instead of killing bacteria, may provide a solution to antibiotic resistance (Hentzer and Givskov, 2003). With the promise of anti-QS compounds, one should be compelled to search for these agents by the most efficient method possible. There have been many ethnobotanically directed searches for agents to treat infection, demonstrating not only the need for these drugs, but also the large number of plants utilized against bacteria (*e.g.*, Cowan, 1999; Camporese *et al.*, 2003; Hernandez *et al.*, 2003). Although this antibacterial effect is important, it is not the only source of a plant's medicinal properties. Shifting the focus from the strictly antibacterial to anti-QS properties of plants may reveal new quorum quenching compounds and provide useful validation for traditional medicinals.

Many plants were reported to have dermatological property. Among which *P. zeylanica* (Plumbaginaceae) is a tropical shrub. It grows wild as a garden plant in eastern, northern and southern India and Ceylon. The roots and leaves of *P. zeylanica* are widely used medicinally in India and China. Traditionally, *P. zeylanica* has been used for the treatment of dermatological disorders including wounds, eczema, scabies, leishmaniasis and leprosy (Abebe and Ayehu, 1993). According to various reported studies, though the root, root barks, and seeds of *P. zeylanica* are used medicinally, the root is the chief source of an acrid crystalline principle called plumbagin; a yellow naphtoquinone pigment, and also characteristic of plants in the tribe Plumbaginaceae including *Plumbago capsensis, P. europea* and *P. rosea*.

The present study was undertaken to investigate the *in vitro* antibiofilm potential of some commonly used medicinal plants against *Staphylococcus aureus* and *Staphylococcus epidermidis* biofilms.

2. Materials and Methods

2.1. Collection of Plant Material

Parts of forty different medicinal plants were collected from Coimbatore. All plants were identified immediately after collection and dried and powdered.

2.2. Test Organisms, Media and Growth Condition

Test organisms such as *Pseudomonas aeroginosa, Salmonella typhi, Shigella* sp., *Escherichia coli, Proteus vulgaris, Enterococcus faecalis, S. aureus* and *S. epidermidis* procured from Institute of Medical Science, PSG Hospitals, Coimbatore, Tamil Nadu, were used during the investigation. Bacterial strains were grown in Luria-Bertani (LB) agar medium at 37° C for 24 hrs.

2.3. Preparation of Crude Extract

To prepare crude extract the dried powder was soaked in 95 per cent ethyl alcohol and water separately for 24 h. The extract was then filtered and stored at –20°C until further analysis. Extracts were tested for microbial contamination at every step of processing by streaking to LB agar plates.

2.4. *In vitro* Antibacterial Assay

The antibacterial activity of crude extract was tested by agar diffusion assay (Barry and Thornsberry, 1985). The plates were incubated at 37°C for 24 h during which activity was evidenced by the presence of zone of inhibition surrounding the well. Each test was repeated three times and the antibacterial activity was expressed as the mean of diameter of the inhibition zones in millimeters (mm).

2.5. Minimum Inhibitory Concentration

The minimum inhibitory concentration of the plant extract against the *S. aureus* and *S. epidermidis* was determined using the agar disc method. Serial dilutions of the plant extracts were prepared to obtain 100-1000 µL. The growth was observed to determine the sensitivity of each organism. The least concentration of the plant extract that had inhibitory effect was taken as the minimum inhibitory concentration (MIC) of the plant extract against such organisms.

2.6. Inhibition of Protease

Protease activity was detected by using skim milk agar plate (Martley *et al.,* 1970). *S. aureus* and *S. epidermidis* was inoculated with Tryptic soy broth (TSB) with ethanolic extract and incubated at 37°C at 180 rpm. Broth was centrifuged at 13000 rpm for 15 min and supernatant was collected and stored at –20°C until further use. Small wells cut in the milk agar plates were filled with 40 µl aliquots of suspensions of staphylococcal cultures and *P. zeylanica* ethanolic mixture and of the respective bacterial supernatants. The digested substrate formed clear areas surrounding the wells. The areas were measured after overnight incubation at 37°C.

2.7. Inhibition of Lipase

Tribuytrin agar medium (Lawrene *et al.,* 1967) was used to determined inhibition of lipase. Medium was poured and microorganisms tested had a lipolytic activity as opaque hallow could be easily observed around the colonies. Inhibition of lipolytic activity was done using tribuytrin agar plate by adding 50 µL of extract with *S. aureus* and *S. epidermidis* to wells grown in the medium suspended with ethanolic extract of *P. zeylanica*. The plates were incubated at 37°C for 48 h. Activity is evaluated by measuring zone of clearance. Lipolytic activity will be observed on a opaque hallow around the colonies.

2.8. Hydrophobicity Index

The effect on Hydrophobicity index of *S. aureus* and *S. epidermidis* were measured by bacterial adhesion to hydrocarbon (BATH) (Zhang and Miller, 1992). Briefly, the bacterial cells grown in TSB broth were resuspended to an OD of 1.0±0.01 at 600 nm. The effect of *P. zeylanica* crude extract on hydrophobicity index was measured by adding the ethanolic extract to bacterial cultures and 1 mL of hexadecane was added to 4mL cell suspension in a test tube and was vortexed for 1 min. The mixture was allowed to settle and separate for 30 min, and the OD of the aqueous phase was measured. The hydrophobicity index (HI) of microbial cells was calculated by the following formula (Serebryakova *et al.*, 2002)

$$\frac{\text{OD Initial} - \text{OD Final}}{\text{OD Initial}} \times 100$$

2.9. Microscopic Method

Biofilm were allowed to grow on glass beads in glass test tubes on TSB supplemented with the 10 µg/mL of *P. zeylanica* extract. Ethanol was added for positive control and was incubated at 37°C for 24 h. After incubation, the glass beads were washed four times with 0.85 per cent normal saline solution (NSS) and each tube was stained with 0.1 per cent crystal violet solution for 10 min. Stained glass beads were observed under microscope. (Lembke *et al.*, 2006).

2.10. Adherent Assay: Tube Method

The method was done following procedure of Christensen *et al.*, 1982. *S. aureus* and *S. epidermidis* were inoculated into 5 mL of TSB in glass tubes in triplicate. Saccharide free basal medium (TSB without glucose) that lacks the substrate for polysaccharide was used as a control. Culture was incubated at 37°C for 20-24 h and the contents were aspirated. One tube was examined unstained and one each stained with crystal violet and trypan blue. Slime positivity was judged by the presence of visible unstained or stained film lining the wall of the tube. Formation of a ring at the liquid air interface was not considered as positive test. Inhibition of glycocalyx production was confirmed by slime negativity.

2.11. Quantification Method

The effect of the *P. zeylanica* crude extract on biofilm formation of *S. aureus* and *S. epidermidis* were investigated by adding the ethanolic extract into glass test tubes containing TSB added with 0.25 per cent glucose. Subsequently, the tubes were supplemented with the bacterial suspension. The bacterial growth was quantified in a spectrophotometer at 415 nm. The tubes containing the media and the etanolic extract of *P. zeylanica* was taken as positive control.

2.12. Effect of *P. zeylanica* Ethanolic Extract against Bacterial Growth Curve

The effect of *P. zeylanica* extract on cell proliferation of *S. aureus* and *S. epidermidis* were determined. Briefly, an overnight culture of *S. aureus* and *S. epidermidis* on TSB medium were diluted to obtain 0.1 OD at 600 nm. Fresh broth were inoculated with these culture and effect of the extract on biofilm formation and growth were monitored at 2 h intervals until a final time point of 24 h.

3. Result and Discussion

3.1. Collection of Plant Material

Based on the medicinal properties from the literature the list of root and rhizome were procured from local market at Madurai. Rest of the samples were collected, shade dried and was used for further analysis. List of the plants used in this study given in Table 20.1.

Table 20.1: List of collected plants.

Sl.No.	Botanical Name	Parts Used	Area of Collection
1.	*Acacia caesia* L.	Bark	Bharathiar University campus
2.	*Acacia nilotica* L.	Bark	Bharathiar University campus
3.	*Acacia sinuate* (Lour.) Metrr	Pods	Simmakal (Madurai)
4.	*Achyranthes aspera* L.	Leaves	Bharathiar University campus
5.	*Acorum calamus* L.	Root	Simmakal (Madurai)
6.	*Aleo vera* L.	Leaf	Kathakinaru (Madurai)
7.	*Alternanthera sessilis* L.	Leaves	Kathakinaru (Madurai)
8.	*Azadirachta indica* A.Juss.	Leaves	Kathakinaru (Madurai)
9.	*Bombax ceiba* L.	Flowers	Bharathiar University campus
10.	*Caesalpinia sappan* L.	Leaves	Bharathiar University campus
11.	*Cassia alata* L.	Leaves	Kathakinaru (Madurai)
12.	*Cassia auriculata* L.	Leaves	Melur (Madurai)
13.	*Cassia fistula* L.	Bark	Bharathiar University campus
14.	*Cassia obtusifolia* L.	Seeds	Bharathiar University campus
15.	*Cassia senna* L.	Leaves	Bharathiar University campus
16.	*Cassia tora* L.	Leaves	Bharathiar University campus
17.	*Centella asiatica* L.	Leaves	Simmakkal (Madurai)
18.	*Cicer arietinum* L.	Seeds	Simmakkal (Madurai)
19.	*Coccinia indica*	Roots, Stem, Leaves	Bharathiar University campus
20.	*Coscinium fenestratum* (Gaerth.) Colebr.	Roots	Simmakkal (Madurai)

Contd...

Table 21.1–Contd...

Sl.No.	Botanical Name	Parts Used	Area of Collection
21.	Crotalaria retusa L.	Leaves	Bharathiar University campus
22.	Curcuma longa L.	Rhizome	Simmakkal (Madurai)
23.	Cymbopogon citrates (DC.) Stapf.	Leaves	Bharathiar University campus
24.	Cyperus rotundus L.	Tubers	Kathakinaru (Madurai)
25.	Diplocyclos palmatus L.	Leaves	Bharathiar University campus
26.	Eclipta prostrate L.	Leaves	Kathakinaru (Madurai)
27.	Eleusine coracana (L.) Gaerth	Grain	Kathakinaru (Madurai)
28.	Embelica ribes Burm.	Leaves	Bharathiar University campus
29.	Ficus religios L.	Leaves	Bharathiar University campus
30.	Foeniculum vulgare Mill.	Fruits	Simmakkal (Madurai)
31.	Hibiseus sabdariffa L.	Leaves	Kathakinaru (Madurai)
32.	Holoptelia integrifolia (Roxb.)	Bark	Bharathiar University campus
33.	Leucas aspera Spreng.	Leaves	Kathakinaru (Madurai)
34.	Murraya koenigii L.	Roots	Kathakinaru (Madurai)
35.	Ocimum americanum L.	Leaves	Kathakinaru (Madurai)
36.	Ocimum basilicum L.	Leaves	Kathakinaru (Madurai)
37.	Ocimum tenuiflorum L.	Leaves	Kathakinaru (Madurai)
38.	Pisum sativum L.	Seeds	Simmakkal (Madurai)
39.	Plectranthus vettiveroides	Roots	Simmakkal (Madurai)
40.	Plumbago zeylanica L.	Roots	Simmakkal (Madurai)

3.2. *In vitro* Antibacterial Assay

Extract of forty medicinal plants were initially tested against test organisms *S. aureus* and *S. epidermidis*. Among the forty plants, ethanolic extracts of 8 plants and both the aqueous as well as ethanolic extracts of 23 plants showed activity against *S. aureus*. Out of the 23 plants zone of inhibition above 10 mm was observed against the ethanolic extract of *Acacia nilotica, Achyranthes aspera, Aleo vera, Alternanthera sessilis, Azadirachta indica, Caesalpinia sappan, Cassia alata, Cassia auriculata, Cassia senna, Curcuma longa, Eclipta prostrate, Foeniculum vulgare, Hibiscus sabdariffa, Murraya koenigii, Ocimum americanum, Ocimum basilicum* and *P. zeylanica*.

3.3. Antimicrobial Activity

Based on the medicinal uses 40 different medicinal plants were collected. Out of 40 plants 19 plants were found to be active against *Staphylococcus aureus* and *S. epidermidis*.

The results obtained showed that ethanolic extracts of *Cassia senna, Cassia alata, Hibiscus sabdariffa, Eclipta prostrate, Alternanthera sessilis* and *Caesalpinia sappan* had maximum inhibitory effects against *Staphylococcus aureus*. It shows that the aqueous extract of *Caesalpinia sappan* (13 mm), *Cassia senna* (10 mm), *Hibiscus sabdariffa* (9 mm),

Ocimum basilicum (9 mm) showed highest inhibitory effects against *S. aureus*. The aqueous extract of *Achyranthes aspera, Acorum calamus, Cassia obtusifolia and Plectranthus vettiveroides* had minimum inhibitory action against *S. aureus*. However, in this study, the ethanolic extract of *Achyranthes aspera, Acacia nilotica, Azadirachta indica, Ocimum americanum* and *Ocimum tenuiflorum* had minimum inhibitory effects on the *S.aureus*. *Acacia caesia, Cassia obtusifolia, Cassia tora, Coccinia indica, Cyperus rotundus* and *Diplocyclos palmatus and Eleusine coracana* had no inhibitory effects against *S. aureus*. *Acacia caesia, Cassia obtusifolia, Cassia tora, Coccinia indica, Cyperus rotundus, Diplocyclos palmatus* and *Pisum sativum* which showed had no inhibitory activity both ethanolic and aquous extract against *S. aureus*.

Inhibition was observed with ethanolic extract of *Caesalpinia sappan* (14 mm), *Cassia alata* (19 mm), *Cassia senna* (15 mm), *Hibiscus sabdariffa* (15 mm) *and Ocimum basilicum* (16 mm) had higher activity against *Staphylococcus epidermidis*. *Centella asiatica* (11 mm), *Curcuma longa* (9 mm), *Ocimum americanum* (13 mm), *Alternanthera sessilis* (13 mm) and *Azadirachta indica* (14 mm) which showed minimum inhibitory action against *S. epidermidis*. In this study aqueous extract of *Cassia alata, Caesalpinia sappan, Cassia senna* and *Ocimum americanum* had good activity against *S. epidermidis*. Ethanolic extract of *Plumbago zeylanica* root had high activity against *S. aureus* (23 mm) and *S. epidermidis* (20 mm), the plant showed the aqueous extract, had highest activity against *S. aureus* (17 mm) and *S. epidermidis* (12 mm). Based on these results *P. zeylanica* was selected for further studies.

3.4. Antimicrobial Activity of *P. zeylanica*

The antimicrobial activity of the ethanolic extract showed different selectivity for each microorganism. The results revealed that ethanolic extract was found to have activity against all tested organisms except *K. pneumoniae* and *Proteus vulgaris*. Ethanolic extract exhibited moderate antibacterial activity against *E. faecalis*, whereas its activity toward the highest inhibition zone was observed against *E. coli* (22 mm), *P. aeruginosa* (19 mm), *S. typhi* (16 mm), crude extract of *P. zeylanica*, which showed the best activity against *Shigella* sp. (24 mm).

3.5. Inhibition of Protease and Lipase

Proteolytic activity of *S. aureus* and *S. epidermidis* were observed by zone of clearance around the colonies. However the inhibition of proteolytic activity was confirmed by the absence of zone of clearance in the ethanolic extract of *P. zeylanica* added plates. In contrast to *S. aureus, S. epidermidis* produces a very limited number of tissue-damaging exoenzymes and toxins. An extracellular metalloprotease of 32 KDa and cysteine protease activity have been reported in *S. epidermidis* (Sloot *et al.*, 1992; Teufel and Gotz, 2000). However, the contribution of these proteases to the pathogenicity of *S. epidermidis* is highly speculative until now and deserves further investigation.

Lipase is an important lipolytic enzyme of *S. aureus* that contributes significantly to the pathogenesis of staphylococcal infection (Brockerhoff and Jensen, 1974). Lipolytic activity of *S. aureus* and *S. epidermidis* were observed by zone of clearance around the colonies. However the inhibition of lipolytic activity was confirmed by reduction in the diameter of zone of clearance in the *P. zeylanica* added wells. Lipase is an important lipolytic enzyme of *S. aureus* that contributes significantly to the pathogenesis of staphylococcal infection (Brockerhoff and Jensen, 1974).

3.6. Effect of *P. zeylanica* on Hydrophobicity Index

The adherence capacity of Staphylococci to host cells depends on the bacterial surface properties like hydrophobicity. Hydrophobicity index is a major determinant of biofilm formation in *S. aureus* and *S. epidermidis*. *P. zeylanica* significantly reduced the cell surface hydrophobicity of *S. aureus* and *S. epidermidis*. The growth of *S. aureus* and *S. epidermidis* cells results in high HI of 79.3 per cent and 94.8 per cent respectively. However, there were significant differences in cells of *S. aureus* and *S. epidermidis* grown in the presence of *P. zeylanica* extract. The HI was observed as 20.39 per cent and 29.95 per cent for *S. aureus* and *S. epidermidis* respectively. As HI and quorum sensing play crucial role in biofilm formation of *S. aureus* and *S. epidermidis*, extracts of *P. zeylanica* acted as inhibitors for these activities. Interestingly, cell viability was higher in the presence of solvents than in single aqueous phase, for the same n-hexane concentration. Viability decreased considerably in the presence of crude extract, cell viability was higher in the presence of solvents than in single aqueous phase, for the same n-hexane concentration. This was observed in the presence of *P. zeylanica* extract. Any disruption of hydrophobic bonds may affect the binding capacity of the bacteria. *S. epidermidis* was reported to be more hydrophobic than *S. aureus*. Being more hydrophobic, they adhere strongly when the biofilms were formed in the presence of ethanol and n-hexane, where as the viability of the rough variant decreased by respectively, when *P. zeylanica* extract was used. Cells of the non-EPS producer variant forming the biofilm responded to the presence of n-hexadecane by increasing the degree of saturation of the fatty acids of the membrane. The degree of saturation increased with increasing number of carbon atoms in the carbon chain of even-numbered n-alkanes.

3.7. Effect of *P. zeylanica* on Dispersion of Biofilm (Microscopic Method)

The anti-biofilm activity of the *P. zeylanica* against *S. aureus* and *S. epidermidis* were observed in light microscopy (Figure 20.1). Anti-biofilm activity of the *P. zeylanica* against *S. aureus* and *S. epidermidis* were significant. The picture shows the effectiveness of the ethanolic extract of *P. zeylanica* eradicating the biofilm formation. The *P. zeylanica* revealed the remarkable ability to disturb biofilm. Analysis of the glass beads using light microscope indicates disruption of biofilm architectures after treatment with the best biofilm inhibitor *P. zeylanica* for a short duration. Increase in the incubation period resulted in the removal of the *S. aureus* and *S. epidermidis* cells from glass beads. These observations justify the real potential of the *P. zeylanica* in disturbing the architectures of *S. aureus* and *S. epidermidis* biofilms.

Adherent Tube Assay

Results showed a positive glycocalyx production in *S. aureus* and *S. epidermidis* whereas, the ethanolic extract of *P. zeylanica* treated shows the negative glycocalyx production in the tube. The slow growth of biofilm bacteria (Costerton *et al.*, 1985;

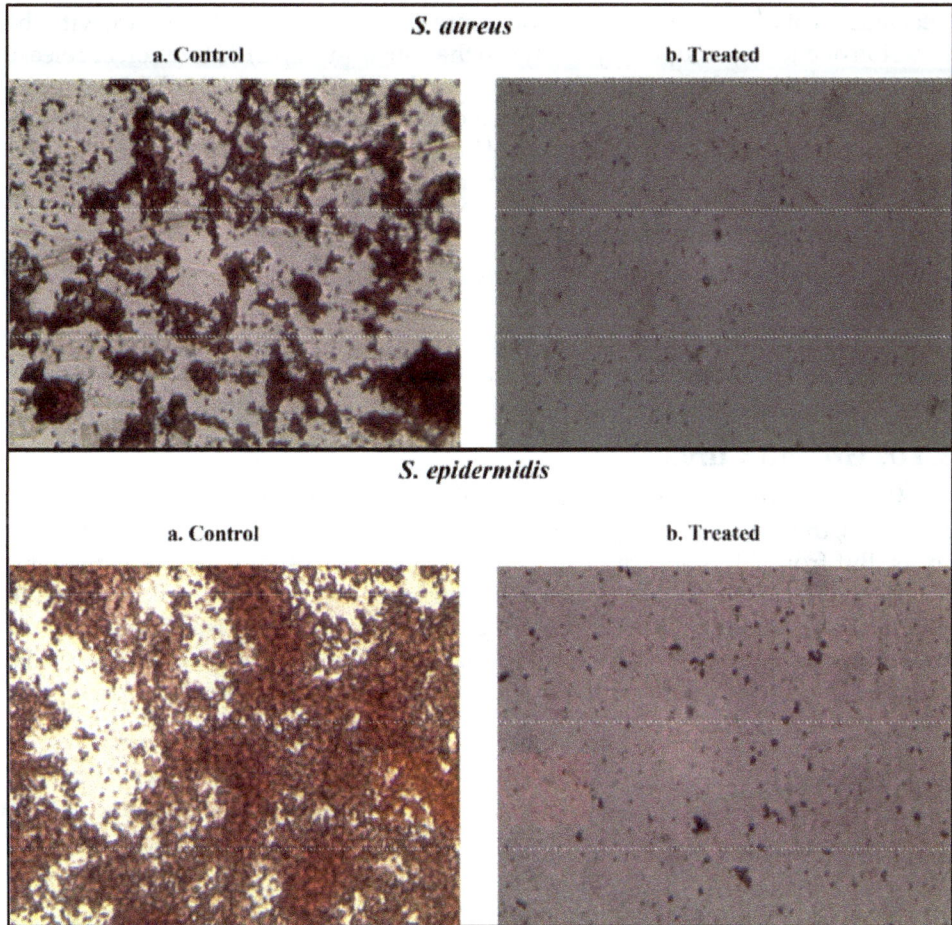

Figure 20.1: Effect of *P. zeylanica* on dispersion of biofilm (microscopic method).

Dickinson and Widmer *et al.*, 1991) and EPS or glycocalyx acting as barrier to the penetration of antibacterial agents (Farber *et al.*, 1990; Evans *et al.*, 1991) are considered responsible for failure of antimicrobial therapy for these infections. Therefore, the removal of glycocalyx matrix in the biofilm structure has the potential of enhancing the antimicrobial therapy against biofilm associated infections (Khardori *et al.*, 1995; Schwank *et al.*, 1998). It is now documented that many chronic infections involve colonization by bacteria growing as adherent biofilms within an extended polysaccharide glycocalyx (Yasuda *et al.*, 1993).

In this study, the ethanolic extract of *P. zeylanica* significantly inhibited the glycocalyx production and adhesion of *S. aureus* and *S. epidermidis* to inert surfaces of glass tubes. These results indicate that low doses of the ethanolic extract of *P. zeylanica* can be used to prevent biofilm formation and associated infections. The high doses of ethanolic extract of *P. zeylanica* may be useful in the treatment of biofilm associated infections with very large inoculum of bacteria in the biofilms. Three possible mechanisms of these effects on the biofilm are (1) Electrostatic interference with the adhesion of bacteria and/or glycocalyx to the substratuml, (2) Activation or release of enzymes to disturb the ESP (glycocalyx) in the biofilms and (3) Inhibition of the formation of new glycocalyx. This shows the *P. zeylanica* crude extract can be used as supplemental material for the treatment of biofilm formation associated with *S. aureus* and *S. epidermidis* infections making a new approach for effective therapy.

3.9. Quantification Method

We quantified the level of *S. aureus* and *S. epidermidis* biofilm treated with *P. zeylanica* at 10 µg/mL concentrations. The level of *S. aureus* and *S. epidermidis* biofilm markedly decreased when treated with the ethanolic extract of *P. zeylanica* (OD 0.870) and (OD 0.624) respectively, whereas control showed high value than the treated tubes (OD 1.924).

3.10. Growth Curve

When the ethanolic extract of *P. zeylanica* was added at the beginning of the cell cycle (0 h), there was no change in the growth of *S. aureus* and *S. epidermidis* for few hours. But from 4 h there was a significant control in the growth of *S. aureus* and *S. epidermidis* in the presence of ethanolic extract of *P. zeylanica* when compared to ethanol and bacterial cultures. From the observed results, it is evident that the inhibition of growth started from 4 h for *S. aureus* and *S. epidermidis*. These results indicates that the ethanolic extract of *P. zeylanica* controls the biofilm formation of *S. aureus* and *S. epidermidis*.

4. Conclusion

To conclude it is evident that the crude extract of *P. zeylanica* disintegrates the architectures of the *S. aureus* and *S. epidermidis* microcolonies, suggesting a possible approach in reducing the development of resistance in sessile cells (which forms dense layer of biofilms) to antibiotics. As a future work the mechanism behind the inhibition by the active principle of *P. zeylanica* ethanolic extract will be studied.

References

Abebe, W., 2003. An overview of herbal supplement utilization with particular emphasis on possible interactions with dental drugs and oral manifestations. *J. Dent. Hyg.*, 77: 37–46.

Adonizio, A.L., Downum, K., Bennett, B.C. and Mathee, K., 2006. Antiquorum sensing activity of medicinal plants in southern Florida. *J. Ethnopharmacol.*, 103: 427–435.

Ahmad, I., Mehmood, Z. and Mohammad, F., 1998. Screening of some Indian medicinal plants for their antimicrobial properties. *J. Ethnopharmacol.*, 62: 183–193.

Barry, A.L. and Thornsberry, C., 1985. Susceptibility tests: Diffusion test procedures. In: *Annual of Clinical Microbiology*, 4th edn., (Ed.) E.H. Lennette. American Society for Microbiology, Washington, DC, U.S.A., pp. 978–987.

Brockerhoff, H. and Jensen, R.G., 1974. *Lipolytic Enzymes*. Academic Press, Inc., New York, pp. 293–324.

Camporese, A., Balick, M.J., Arvigo, R., Esposito, R.G., Morsellino, N., Simone, F.D. and Tubaro, A., 2003. Screening of anti-bacterial activity of medicinal plants from Belize (Central America). *J. Ethnopharmacol.*, 87: 103–107.

Cheristensen, G.D., Simpson, W.A., Bisno, A.L. and Beachey, E.H., 1982. Adhesion of slime-producing strain of *Staphylococcus epidermidis* to smooth surfaces. *Infect. Immun.*, 37(1): 318–326.

Costerton, J.W., Nickel, J.C. and Marrie, T.J., 1985. The role of the bacterial glycocalyx and the biofilm mode of growth in bacterial pothogenesis. *Roche Semin. Bacteriol.*, 2: 1–25.

Costerton, J.W., Cheng, K.J., Geesey, G.G., Ladd, T.I., Nickel, J.C., Dasgupta, M. and Marrie, T.J., 1987. Bacterial biofilms in nature and disease. *Annu. Rev. Microbiol.*, 41: 435.

Cowan, M.M., 1999. Plant products as antimicrobial agents. *Clin. Microbiol. Rev.*, 12: 564–582.

Cragg, G.M., Newman, D.J. and Snader, K.M., 1997. Natural products in drug discovery and development. *J. Nat. Prod.*, 60: 52–60.

Cumberbatch, A., 2002. Characterization of the anti-quorum sensing activity exhibited by marine macroalgae of South Florida. In: *Undergraduate Honors Thesis*. Mathee, K. (Mentor), Flordia International University.

Dickinson, G.M. and Bisno, A.L. 1989. Infections associated with indwelling devices; Concepts of pathogenesis. *Antimicrob. Agents Chemother.*, 33: 598–601.

Dong, Y.H. and Zhang, L.H., 2005. Quorum sensing and quorum quenching enzymes. *J. Microbiol.*, 43: 101–109.

Evans, D.J.K., Brown, M.R., Asllison, D.G. and Gilbert, P., 1991. Susceptibility of *Pseudomonas aeruginosa* and *Escherichia coli* biofilms toward ciprofloxacin: Effect of specific growth rate. *J. Antimicrob. Chemother.*, 27: 177–184.

Farber, B.F., Kaplan, M.H. and Clogston, A.G., 1990. *Staphylococcus epidermidis* extracted slime inhibits the antimicrobial action of glycopeptide antibiotics. *J. Infect. Dis.*, 161: 37–40.

Gao, M., Teplitski, M., Robinson, J.B. and Bauer, W.D., 2003. Production of substances by *Medicago truncatula* that affect bacterial quorum sensing. *Am. Phytopathol. Soc.*, 16: 827–834.

Gotz, F. and Peters, G., 2000. Colonization of medical devices by coagulase-negative staphylococci. In: *Infections Associated with Indwelling Medical Devices*, (Eds.) F.A. Waldvogel and A.L. Bisno. American Society for Microbiology Press, Washington, DC, pp. 55–88.

Hentzer, M. and Givskov, M., 2003. Pharmacological inhibition of quorum sensing for the treatment of chronic bacterial infections. *J. Clin. Invest.*, 112: 1300–1307.

Hentzer, M., Wu, H., Anderson, J.B., Riedel, K., Rasmussen, T.B., Bagge, N., Kumar, N., Schembri, M.A., Song, Z., Kristofferson, P., Manefield, M., Costerton, J.W., Molin, S., Eberl, L., Steinberg, P., Kjelleberg, S., Høiby, N. and Givskov, M., 2003.

Attenuation of *Pseudomonas aeruginosa* virulence by quorum sensing inhibitors. *EMBO J.*, 22: 3803–3815.

Khardori, N., Yassein, M. and Wilson, K., 1995. Tolerance of *Staphylococcus epidermidis* grown from indwelling vascular catheters to antimicrobial agents. *J. Indus. Microbiol.*, 15: 148–151.

Kong, K.F., Vuong, C. and Otto, M., 2006. *Staphylococcus* quorum sensing in biofilm formation and infection. *Int. J. Med. Microbiol.*, 296: 133–139.

Lawrence, R.C., Fryer, T.F. and Reiter, B., 1967. Rapid method for the quantitative estimation of microbial lipases. Nature, 213: 1264–1265.

Lembke, C., Podbielski, A., Hidalgo-Grass, C., Jonas, E., Hanski, L. and Kreikemeyer, B., 2006. Characterization of biofilm formation by clinically relevant serotypes of group A *Streptococci. Appl. Environ. Microbiol.*, 72: 2864–2875.

Manefield, M., De Nys, R., Kumar, N., Read, R., Givskov, M., Steinberg, P. and Kjelleberg, S., 1999. Evidence that halogenated furanones from *Delisea pulchra* inhibit acylated homoserine lactone (AHL)–mediated gene expression by displacing the AHL signal from its receptor protein. Microbiology, 145: 283–291.

Martley, F.G., Jarvis, W., Bacon, D.F. and Lawrence, R.C., 1970. Typing of coagulase-positive *Staphylococci* by proteolytic activity on buffered caseinate agar with special reference to bacteriophage nontypable strains. *Infect. Immun.*, 4: 439–442.

Otto, M., 2004. Quorum-sensing control in Staphylococci: A target for antimicrobial drug therapy? *FEMS Microbiol. Lett.*, 241: 135–141.

Schwank, S., Rajacic, Z., Zimmerli, W. and Blasser, J., 1998. Impact of bacterial biofilm formation on *in vitro* and *in vivo* activities of antibiotics. *Antimicrob. Agents Chemother.*, 42(4): 895–899.

Serebryakova, E.V., Darmov, I.V., Medvedev, N.P., Alekseev, S.M. and Rybak, S.I., 2002. Evaluation of the hydrophobicity of bacterial cells by measuring their adherence to chloroform drops. *Microbiology*, 71: 202–204.

Sloot, N., Thomas, M., Marre, R. and Gatermann, S., 1992. Purification and characterization of elastase from *Staphylococcus epidermidis. J. Bacteriol.*, 37: 201–205.

Teplitski, M., Robinson, J.B. and Bauer, W.D., 2000. Plants secrete substances that mimic bacterial N-acyl homoserine lactone signal activities and affect population density-dependent behaviors in associated bacteria. *Mole. Plant Microbe Interact.*, 13: 637–648.

Teufel, P. and Gotz, F., 1993. Characterization of an extracellular metalloprotease with elastase activity from *Staphylococcus epidermidis. J. Bacteriol.*, 175: 4218–4224.

Widmer, A.F., Weistner, A., Frei, R. and Zimmerli, W., 1991. Killing of non growing and adherent *Escherichia coli* determines drug efficacy in device-related infections. *Antimicrob. Agents Chemother.*, 35: 741–746.

Xu, L., Li, H., Vuong, C., Vadyvaloo, V., Wang, J., Yao, Y., Otto, M. and Gao, Q., 2006. Role of the luxS quorum-sensing system in biofilm formation and virulence of *Staphylococcus epidermidis. Infect. Immun.,* 74: 488–496.

Yasuda, H., Ajiki, Y., Tetsufumi, K., Harumi, K. and Yokoya, T., 1993. Interaction between bifilms formed by *Pseudomonas aeruginosa* and clarithromycin. *Antimicrob. Agents Chemother.,* 37: 1749–1755.

You, J., Xue, X., Cao, L., Lu, X., Wang, J., Zhang, L. and Zhou, S.V., 2007. Inhibition of Vibrio biofilm formation by a marine actinomycetes strain A66. *Appl. Microbiol. Biotechnol.,* 76: 1137–1144.

Zhang, Y. and Miller, R.M., 1992. Enhanced octadecane dispersion and biodegradation by a *Pseudomonas rhamnolipid* surfactant (Biosurfactant). *Appl. Environ. Microbiol.,* 58: 3276–3282.

Scientific Basis of Herbal Medicine (2013)   Pages 191–194
Editor: Dr. Parimelazhagan Thangaraj
Published by: DAYA PUBLISHING HOUSE, NEW DELHI

Chapter 21

Utilization of Herbals in Traditional Medicinal System of Ethiopia

R. Hiranmai Yadav¹ and B. Vijayakumari²

¹School of Natural Resource Management and Environmental Sciences,
College of Agriculture, Haramaya University, Dire Dawa, Ethiopia
²Department of Botany, Avinashilingam University,
Coimbatore – 641 043, Tamil Nadu, India

1. Introduction

The World Health Organization (WHO) defines traditional medicine as health practices, approaches, knowledge and belief incorporating plant, animal and mineral based medicines, spiritual therapies, manual techniques and exercises, applied singularly or in combination to treat, diagnose and prevent illness and maintain well being (WHO, 2001). Some of the primary health needs in African, Asian and Latin American countries are met by traditional medicine. About 80 per cent of population in African countries uses traditional medicine for primary health care. The traditional medicine has maintained its popularity in developing countries (Papadopoulos *et al.*, 2002).

Ethiopia have a long history of traditional medicine which is diverse among different cultures. The healing in this system is concerned with protection and promotion of human physical, spiritual, social, mental and material well being. The system is a holistic approach. The traditional healing knowledge is passed orally within certain families or social groups. The healers obtain the plants, animals and minerals used in the treatment from the natural resources. They are prepared in different forms and prescribed in a non formulated form and administered either by topical, oral or respiratory routes. There are antidotes in case of side effects or discomfort by the drugs (Beshaw, 1991; WHO, 2005).

The medico religious manuscript in the country confirms the long history of Ethiopia using traditional system of using medicinal plants (Kibebew, 2001). The plant remedies are found to be sole source of medication for about 80 per cent of population in the country (Abebe, 2001).There is also a loss of medicinal plants due to natural and anthropogenic reasons.

The present study was an attempt to gather knowledge regarding the different herbal treatment systems and plants utilized for the treatment of different ailments.

2. Materials and Methods

The traditional system of medicine is practiced for generations and the system is found to be more in harmony with nature.

A survey was conducted to develop knowledge about the traditional practices in the area. Different ethnic groups have distinct and unique traditional practices. There are different names given to the healers (woghesha for bone setter, awalag for midwife, debtera for religious man).The healers are mainly males who gained their knowledge from relatives or transferred from their parents orally. The healers treat both humans and veterinary diseases.

The diverse topography of the country provides habitat for more than 7,000 higher plant species which has a diverse endemic flora with therapeutic values. Few of these plants that are commonly used for fumigation, vermifuge, pain relief, skin infection and a variety of common ailments are sold in markets along with food and spices.

3. Results and Discussion

The current survey showed that most of the practitioners are males with almost 98 per cent of them are traditional healers. The healers reveal the medicinal plants or the treatment methods only to their predecessors who will take up the role of healers in future.

The plants used for ailments include mostly herbs, shrubs, trees, climbers, lianas, epiphytes etc. These are collected from wild by the healers. They identify the plants and collect and use it in raw or processed forms. They mostly collect leaves and roots which are used in larger quantities. In addition they also use fruits, bark, stem, seed and whole plant. The treatment system uses both dry and fresh remedies.

The traditional medicine are prepared mainly by pounding, powdering, smashing, chewing, rubbing, squeezing, cream, crushing, steam bath, dry bath(Smoke), burning and rinsing.

The major routes of administration in the study area are oral, dermal, nasal, anal, auricular and optical. The methods include drinking, swallowing, sniffed, eating, chewing, dry bath (smoke), creamed, steam bath, washing, bandaging etc. The plants or parts may be used singly or in combination of different plants. The locals prefer fresh forms and consider it to be effective compared to dry forms.

The healers are approached for both human and veterinary treatment. They always have secret herbs and ingredients (Mesfin and Obsa, 1994). Ethiopian traditional medical system is characterized by variation which is shaped by ecological diversities of the country, socio ecological back ground of different ethnic groups and historical developments that could be related with migration. The present system hails from three main medical sub-systems that existed in the country namely, Cushitic folk medicine, Arabic medicine and Amhara medicine. The history, perceptions about health and healers in these system is distinct. It is an amalgamation of beliefs and empirical practices (Abebe and Aychu, 1993; Pankhurst, 1990; Slikkerveer,1990).The indigenous knowledge is still conveyed verbally and includes superstitious practices and beliefs(Getachew Addis *et al.*, 2002).The traditional medicine system also reflects the diversity in the culture which is a holistic approach to bring about physical, mental and social well being (Kebede *et al.*, 2006).

Table 21.1: Few plants regularly used.

Sl.No.	Plant Name	Part Used
1.	*Acacia mellifera*	leaf
2.	*Acalypha nilotica*	leaf
3.	*Aristolochia bracteolate*	*leaf*
4.	*Balanites aegyptica*	leaf
5.	*Balanites rotundifolia*	whole plant
6.	*Boscia coriacea*	whole plant
7.	*Coffea arabica*	seed
8.	*Commicarpus plumbagineus*	root
9.	*Heliotropium rariflorum*	whole plant
10.	*Indigofera* sp	Whole plants
11.	*Lycium shawii*	root
12.	*Salvadora persica*	root
13.	*Senna italica*	Whole plant
14.	*Sericomposis pallida*	root
15.	*Solanum coagulans*	seed
16.	*Solanum coagulans*	whole plant
17.	*Solanum incanum*	fruit
18.	*Xanthium strumarium*	leaf
19.	*Zaleya pentandra*	leaf

4. Conclusion

Considering the strict secrecy maintained by the healers the treatment becomes more reliable and less susceptible to distortion. The use of traditional medicine is widespread in urban and rural community which shows the reliability and cultural acceptability. The traditional knowledge is also associated with plant resources that have food values and general utility along with medicinal properties. The indigenous knowledge is developed in people by living in harmony with nature. There is considerable variation in traditional knowledge of people about herbs that vary with different communities and also socio demographic factors. The resources are utilized by the different generations in varying forms. The utilization of natural resources in a harmonious way will be beneficial for preserving the knowledge and resources for future. The enormous demand of the natural flora leads to indiscriminate harvesting thereby resulting in loss of flora. To ensure a continuous supply of these valuable herbs there should be alternative development of the plants by cultivation.

References

Abebe, D., 1998. The role of medicinal plants in healthcare coverage of Ethiopia, the possible benefits of integration. In: *Conservation and Sustainable Use of Medicinal Plants in Ethiopia*, (Ed.) M. Zewdu. Proceedings of the National Workshop, 28 April–01 May. Addis Ababa. Edited by A. Demissie. Institute of Biodiversity Conservation and Research, 2001, pp. 6–21.

Abebe, D. and Aychu, A., 1993. *Medicinal Plants and Enigmatic Health Practices of Northern Ethiopia*. Addis Ababa, B.S.P.E.

Beshaw, M., 1991. Promoting traditional medicine in Ethiopia: A brief historical overview of government policy. *Soc. Sci. and Med*, 33: 193–200.

Getachew, Addis, Dawit, Abebe, Timotewos, Genebo and Kelbessa, Urga, 2002. Perceptions and practices of modern and traditional health practitioners about traditional medicine in Shirka District, Arsi zone, Ethiopia. *Ethiop. J. Health Dev.*, 16: 19–29.

Kebede Deribe Kassaye, Alemayehu Amberbir, Binyam Getachew and Yunis Mussema, 2006. A historical overview of traditional medicine practices and policy in Ethiopia. *Ethiop. J. Health Dev.*, 20: 127–134.

Kibebew, F., 2001. The status and availability of oral and written knowledge on traditional health care on traditional health care in Ethiopia. In: *Conservation and Sustainable Use of Medicinal Plants in Ethiopia*, (Ed.) M. Zewdu. Proceedings of the National Workshop, 28 April–01 May. Addis Ababa. Edited by A. Demissie. Institute of Biodiversity Conservation and Research, 2001, pp. 107–119.

Mesfin, T. and Obsa, T., 1994. Ethiopian traditional veterinary practices and their possible contribution to animal production and management. *Rev. Sci. Tech. Off. int. Epiz.*, 13: 417–424.

Pankhurst, R., 1990. *An Introduction to Medical History of Ethiopia*. Trenton, New Jersey, The Red Sea Perss, Inc.

Papadopoulos, R., Lay, M. and Gebrehiwot, A., 2002. Cultural snapshots: A guide to Ethiopian refugees for health care workers. Research Center for Trans-cultural Studies in Health, Middlesex University, London UK. N 14 4YZ, May 2002. Available at: www.mdx.ac.uk/www/rctsh/embrace.htm.

Slikkerveer, I.J., 1990. *Plural Medical Systems in the Horn of Africa: The Legacy of Sheik Hippocrates*. Kegan Paul International, London and New York.

WHO, 2001. *Legal Status of Traditional Medicine and Complementary/Alternative Medicine: A World Wide Review*. Geneva.

WHO, 2005. *National Policy on Traditional Medicine and Regulation of Herbal Medicines*. Report of a WHO Global Survey, Geneva, Switzerland.

Scientific Basis of Herbal Medicine (2013)
Editor: Dr. Parimelazhagan Thangaraj
Published by: DAYA PUBLISHING HOUSE, NEW DELHI

Pages 195–200

Chapter 22

Studies on Antibacterial Activity of Various Extracts of the Leaves of the Medicinal Climber, *Solena amplexicaulis* (Lam.) Gandhi.

Karthika Krishna Moorthy and Subramanian Paulsamy

Department of Botany,
Kongunadu Arts and Science College,
Coimbatore –641 029, Tamil Nadu, India

1. Introduction

Nature has been a source of medicinal agents for thousands of years and since the beginning of mankind (Kafaru, 1994). Plants with their wide variety of chemical constituents offer a promising source of new antimicrobial agents with general as well as specific activity (Evans, 1996). The herbal medicines are being used since time immemorial for the treatment of uncountable diseases (Gordon and David, 2001). In recent years, multiple drug resistance in human pathogenic microorganisms has developed due to indiscriminate use of commercial antimicrobial drugs for the treatment of infectious diseases. The cost of production of synthetic drugs is also high and they produce adverse effects compared to plant derived drugs. Hence much attention has been paid recently to the biologically active compounds derived from plants used in herbal medicine (Firas and Bayati, 2008; Shai *et al.*, 2008). There are several reports on the presence of antimicrobial compounds in various plants (Ahmad *et al.*, 1998; Nair *et al.*, 2005; Prusti *et al.*, 2008).

Solena amplexicaulis (Lam.) Gandhi. is a perennial dioeceous climber commonly known as creeping cucumber belonging to Cucurbitaceae family is widely distributed throughout tropical Asia. The whole plant is edible and potential source of natural antioxidant (Venkateshwarlu *et al.*, 2011; Karthika *et al.*, 2012). The tubers, leaves and seeds of the plant are extensively used in traditional system for various ailments like spermatorrrhoea, thermogenic, appetizer, cardiotonic, diuretics, haemorrhoids and invigorating (Kritchevsky, 1978). The leaves of this plant have good anti-inflammatory activity. Hence it is recommended for the treatment of inflammation, skin lesions and skin diseases (Arun *et al.*, 2011). Root is stimulant and purgative (Goldin Quadros, 2009).

Considering the indigenous uses of the plant, the present investigation was taken up with an objective to evaluate the antibacterial potential against certain human pathogenic bacteria.

2. Materials and Methods

2.1. Plant Material

The leaves of *S. amplexicaulis* were collected from Madukkarai, Coimbatore district, Tamil Nadu. Collected plant materials were washed thoroughly in tap water, shade dried and then homogenized to fine powder and stored in air tight bottles.

2.2. Preparation of Extracts

About 50g coarsely powdered plant material (50 g/250 ml) was extracted in the soxhlet extractor for 8 to 10 hours, sequentially with hexane, benzene, chloroform, methanol and water. Then the each extracts were evaporated to dryness.

2.3. Bacterial Strains

In vitro antibacterial activity was examined for the crude extracts of leaves of the plant, against 9 bacterial species which include the Gram positive strains *viz.*, *Streptococcus faecalis*, *S. pyogenes*, *Enterococcus faecalis* and *Bacillus subtilis* and Gram negative strains *viz.*, *Salmonella paratyphi* A, *S. paratyphi* B, *Klebsiella pneumonia*, *Serratia marcescens* and *Pseudomonas aeruginosa*.

All these bacterial strains were obtained from the Department of Microbiology, Tamil Nadu Agricultural University, Coimbatore. All the bacteria were maintained at 4°C on nutrient agar slants for further use.

2.4. Bacterial Susceptibility Resting

An inoculum of each of the pathogenic bacterial strains was suspended in 5ml of nutrient broth and incubated at 37°C for 18 hrs. Bioassay was carried out by agar well diffusion method (Perez *et al.*, 1990; Murray *et al.*, 1995; Olurinola, 1996). Inoculum was spread over Muller – Hinton agar medium with sterile glass spreader. A well of 6mm diameter was made using a sterile cork borer and filled with 50 µl of different extracts by using micropipette in each well in aseptic condition. The plates were kept at room temperature for absorption of extract in the medium and further incubated in a incubator at 37°C for 24 hrs. The antibacterial activity was evaluated by measuring the diameter of inhibition zone. Amphicillin was used as positive control (50µl/ml) and DMSO (Dimethyl sulphoxide) as a negative control (50 µl).

2.5. Statistical Analysis

All the analysis was done in triplicate and results were expressed as mean±SD. The data were subjected to one way analysis of variance (ANOVA) and the significance of the difference between mean was determined by Duncan's Multiple Range Test

Table 22.1: *In vitro* antibacterial activity of *Solena amplexicaulis* extracts by agar well diffusion method.

Sl.No.	Name of the Bacteria	Diameter of Inhibition Zone (mm)					
		Control*	Hexane	Benzene	Chloroform	Methanol	Water
	Gram Positive Bacteria						
1.	Streptococcus faecalis	23.67±1.15ᵃ	–	11.33±3.51ᵇ	13.33±3.79ᶜ	12.00±4.36ᵇᶜ	–
2.	S. pyogenes	19.00±1.00ᵃ	6.50±1.73ᵇ	11.67±4.51ᶜ	17.00±7.94ᵃᶜ	9.64±3.51ᵇᶜ	–
3.	Enterococcus faecalis	19.00±1.73ᵃ	–	9.67±5.03ᵇ	16.33±5.13ᵃᶜ	14.00±6.56ᶜ	–
4.	Bacillus subtilis	21.67±5.13ᵃ	–	9.67±2.89ᵇ	12.67±4.62ᶜ	7.33±0.58ᵇᵈ	–
	Gram Negative Bacteria						
5.	Salmonella paratyphi A	22.33±2.52ᵃ	9.33±5.13ᵇ	21.00±1.00ᵃ	27.00±1.00ᶜ	11.33±2.89ᵇᵈ	8.00±5.20ᵇ
6.	S. paratyphi B	25.33±6.81ᵃ	7.33±2.52ᵇ	18.00±4.36ᶜ	23.67±3.51ᵈ	10.33±1.51ᵉ	–
7.	Klebsiella pneumonia	14.33±6.03ᵃ	–	10.33±4.73ᵇ	11.33±2.89ᶜ	11.00±2.65ᶜ	–
8.	Serratia marcescens	16.67±3.5ᵃ	–	7.00±1.00ᵇ	14.33±2.08ᶜ	9.00±1.00ᵈ	–
9.	Pseudomonas aeruginosa	17.00±1.73ᵃ	8.33±4.16ᵇ	15.67±5.85ᶜ	18.00±4.00ᵃ	13.00±5.20ᵃᵈ	–

'*' Amphicillin, '-' indicates no activity.

Values were performed in triplicates and represented as mean±SD.

Mean values followed by different superscript in a column are significantly different (P<0.05).

with significance level, P<0.05. ANOVA was performed using the statistical software SPSS (SPSS Inc. Chicago, USA).

3. Results and Discussion

Since multidrug resistance of microorganisms is a major medical concern, screening of natural products in a search for new antimicrobial agents that would be active against these microorganisms is the need of the hour (Zyoda and Porter, 2001). The results of the experiments for the antibacterial activity of the leaves of the study species, *S. amplexicaulis* is given in Table 22.1. In the present study, out of 5 solvents used, the extract obtained from benzene, chloroform and methanol exhibited higher activity and that obtained from hexane and water showed lower activity. The benzene, chloroform and methanol extracts were able to inhibit the growth of all the pathogenic bacteria. This indicates that the active principle compound that inhibits the growth of susceptible bacteria may dissolve better in alcoholic solvents than in water (Marjorie, 1999). Highest activity was by the chloroform and benzene extracts respectively performed against *Salmonella paratyphi* A (27 mm and 21 mm) and *S. paratyphi* B (23 mm and 18 mm). This indicates that this plant can be used to cure the paratyphoid fever, which is caused by any strains of *Salmonella* (Frey and Rebecca, 1999). The methanol extract was determined to be active against *Enterococcus faecalis* (14mm). Gram negative bacteria are more sensitive than that of the Gram positive bacteria to this extract. Among the Gram negative bacteria *Klebsiella pneumoniae* and *Serratia marcescens* were more resistant to these extracts. Gram negative bacteria were frequently reported to have developed multidrug resistant due to their outer membrane which acts as a barrier to many environmental substances including antibiotics (Johansson *et al.*, 2011; Johnson *et al.*, 2011; Ramakant *et al.*, 2011; Tortora *et al.*, 2001). High antibacterial effects in organic extracts for certain Cucurbitaceae members were already well documented (Mustafa *et al.*, 2009; Senthil Kumar and Kamaraj, 2011). As expected, the standard drug Amphicillin showed high degree of inhibition against, *Streptococcus faecalis, Bacillus subtilis, Salmonella paratyphi* A and *S. paratyphi* B. Negative control showed no formation of zone of inhibition. The overall results of the study revealed that the crude extract of the study plant contain certain constituents with significant antibacterial property. The present investigation confirmed the therapeutic potency of *S. amplexicaulis* and justified the traditional use of this species.

4. Conclusion

The investigation reports that the species *S. amplexicaulis* is active against tested pathogenic bacteria which confirms the traditional use of this species for health care system in sourthern India. Further investigations are necessary for the isolation of novel lead compounds. So as to use this plant in pharmacological industries.

Acknowledgement

The authors are gratefully acknowledging University Grants Commission, New Delhi for their financial assistance to carry out the work.

References

Ahmad, I., Mehmood, Z. and Mehmood, I., 1998. Screening of some Indian medicinal plants for their antimicrobial properties. *Journal of Ethnopharmacology*, 62: 183–193.

Arun, Ch., Satheesh Kumar, R., Srinu, S., Lal Babu, G., Raghavendra Kumar, G. and Amos Babu, J., 2011. Anti-inflammatory activity of aqueous extract of leaves of *Solena amplexicaulis. International Journal of Research in Pharmaceutical and Biomedical Sciences*, 2(4): 1617–1619.

Evans, W.C., 1996. *Trease and Evans Pharmacognosy*, 14th edn. WB Saunders Company Ltd., p. 290.

Firas, A. and Bayati, A., 2008. Synergistic antibacterial activity between *Thymus vulgaris* and *Pimpinella anisum* essential oils and methanol extracts. *Journal of Ethnopharmacology*, 116: 403–406.

Frey, Rebecca, J., 1999. Paratyphoid Fever. *Encyclopedia of Medicine.*

Goldin Quadros, 2009. Report of the 'Study of the biodiversity of Indian Institute of Technology Bombay Campus' by World Wide Fund for Nature–India. Maharashtra State Office, Mumbai.

Gordon, M.C. and David, J.N., 2001. Nature product drug discorvery in the next millennium. *Pharmaceutical Biology*, 139: 8–17.

Johansson, M., Phuong, D.M., Walther, S.M. and Hanburger, H., 2011. Need for improved antimicrobial and infection control stewardship in Vietnamese intensive care units. *Tropical Medicine and International Health*, 16(6): 737–743.

Johnson, M., Wesely, E.G., Kavitha, M.S. and Uma, V., 2011. Antibacterial activity of leaves and inter-nodal callus extracts of *Mentha arvensis* L. *Asian pacific Journal of Tropical Medicine* 4(3): 196–200.

Kafaru, E., 1994. *Essential Pharmacology*. Elizabeth Kafaru Publishers, Lagos, Nigeria, Immense Help Formative Workshop, pp. 11–14.

Karthika, K., Paulsamy, S. and Jamuna, S., 2012. Evaluation of *in vitro* antioxidant potential of methanolic leaf and stem extracts of *Solena amplexicaulis* (Lam.) Gandhi. *Journal of Chemical and Pharmaceutical Research*, 4(6): 3254–3258.

Kritchevsky, D., 1978. Fiber, lipids and atherosclerosis. *American Journal of Clinical Nutrition* 31S, 65–74.

Marjorie, M.C., 1999. Plant product as antimicrobial agents. *Clinical Microbiology Reviews*, 12: 564–582.

Murray, P.R., Baron, E.J., Pfaller, M.A., Tenover, F.C. and Yolken, H.R., 1995. *Manual of Clinical Microbiology*, 6th edn. ASM Press, Washington, DC, pp. 15–18.

Mustafa, O., Dilek, O. and Fatih, K., 2009. Activity of some plant extracts against multi-drug resistant human pathogen. *Iranian Journal of Pharmaceutical Research* 8(4): 293–300.

Nair, R., Kalariya, T. and Sumitra Chanda., 2005. Antibacterial activity of some selected Indian medicinal flora. *Turkish Journal of Biology* 29: 41–47.

Olurinola, P.F., 1996. *A Laboratory Maual of Pharmaceutical Microbiology*. Idu, Abuja, Nigeria, pp. 69–105.

Perez, C., Paul, M. and Bazerque, P., 1990. An antibiotic assay by agar well diffusion method. *Acta Biol. Med. Exp.*, 15: 113–115.

Prusti, A., Mishra, S.R., Sahoo, S. and Mishra, S.K., 2008. Antibacterial activity of some selected Indian medicinal plants. *Ethnobotanical Leaflets*, 12: 227–230.

Ramakant, P., Verma, A.K., Misra, R., Prasad, K.N., Chand, G., Mishra, A., Agarwal, G., Agarwal, A. and Mishra, S.K., 2011. Changing microbiological profile of pathogenic bacteria in diabetic food infections: Time for a rethink on which empirical therapy to choose? *Diabetologia*, 54(1): 58–64.

Senthil Kumar, S. and Kamaraj, M., 2011. Antimicrobial activity of *Cucumis anguria* L. by agar well diffusion method. *Botany Research International*, 4(2): 41–42.

Shai, L.J., McGaw, M.A., Aderogbaa, L.K. and Eloff, J.N., 2008. Four pentacyclic triterpenoids with antifungal and antibacterial activity from *Curtisia dentate* (Burm.f.) leaves. *Journal of Ethanopharmacology* 119: 238–244.

Tortora, G.J., Funke, B.R. and Case, C.L., 2001. *Microbiology: An Introduction*. Benjamin Cummings, San Francisco, p. 88.

Venkateshwarlu, E., Raghuram Reddy, A., Goverdhan, P., Swapna Rani, K. and Jayapal Reddy, G., 2011. *In vitro* and *In vivo* antioxidant activity of methanolic extract of *Solena amplexicaulis* (Whole plant). *International Journal of Pharmacy and Biological Science* 1(4): 522–533.

Zyoda, J.R. and Porter, J.R., 2001. A convenient microdilution method for screening natural products against bacteria and fungi. *Pharmaceutical Biology*, 39(3): 221–225.

Scientific Basis of Herbal Medicine (2013)   *Pages* **201–205**
Editor: **Dr. Parimelazhagan Thangaraj**
Published by: **DAYA PUBLISHING HOUSE, NEW DELHI**

Chapter 23

Thrombolytic Activity and Cytotoxicity of Aqueous Extract of Fresh and Dry *Aloe vera*

Ramakrishnan Pappa Ammal,
Amirthalingam Pushpa and Kenthoraj Raju Asha

Department of Biochemistry,
Avinashilingam Institute for Home Science and Higher Education for Women,
Coimbatore – 641 043, Tamil Nadu, India

1. Introduction

A blood clot (thrombus) developed in the circulatory system due to the failure of hemostasis causes vascular blockage and while recovering leads to serious consequences in atherothrombotic diseases such as acute myocardial or cerebral infarction, at times leading to death. All available thrombolytic agents still have significant shortcomings, including the need for large doses to be maximally effective, limited fibrin specificity and bleeding tendency. Due to the shortcomings of the available thrombolytic drugs, attempts are underway to develop improved recombinant variants of these drugs (Anwar *et al.*, 2011).

Platelets play a central role in the process of blood clotting and are regarded as key regulators of both hemostasis and pathogenesis of cardiovascular disease. Mode of action and efficacy of the current anti- platelet agents are still clinically doubtful and they are associated with the risk of severe or fatal bleeding. This has fuelled the search for alternative medicine (Mosa *et al.*, 2011).

The herbal preparation of gel of *Aloe vera* has been used since ancient times for the treatment of several diseases like stomach disorders and intestinal disorders including constipation, hemorrhoids, colitis and colon problems. However, the use of this plant as thrombolytic agent has not been reported so far. The present study aims to find out the thrombolytic and cytotoxic properties of fresh and dry *Aloe vera* gel.

2. Materials and Methods

2.1. Selection of the Plant

Aloe vera belongs to the family of Asphodelaceae. It was reported to have various pharmacologic properties, specifically to promote wound, burn and frost bite healing, in addition to having anti-inflammatory, immunomodulatory and gastroprotective properties. Moreover, some specially prepared *Aloe vera* extracts possess many biological activities such as hypoglycemic, hypolipidemic, antifungal, anticancer, antioxidant and immunoprotective (Shemy *et al.*, 2010). Fresh *Aloe vera* leaves were collected from Avinashilingam University Herbal garden. Dry *Aloe vera* gel was purchased from local market.

2.2. Preparation of Plant Extract

2.2.1. Fresh *Aloe vera* Gel Extract

2g of fresh *Aloe vera* gel was weighed and homogenized with mortar and pestle and filtered through Whatman No. 1 filter paper and the clear filtrate was used for analysis.

2.2.2. Dry *Aloe vera* Gel Extract

About 2 g of dry gel was weighed and dissolved in distilled water and filtered through whatman No. 1 filter paper and clear filtrate was used for analysis.

2.3. Assessment of Thrombolytic Activity

Venous blood was drawn from the healthy volunteers. 500 µl of blood was transferred to preweighed micro centrifuge tubes to form clots.

2.4. Positive Control

The commercially available clavix (75 mg) was dissolved in 1.0 ml of distilled water and the suspension was used as a positive control.

2.5. Study Design

Venous blood drawn from healthy volunteers (n=20) was transferred in different preweighed sterile micro centrifuge tubes (500 µl/ml). After clot formation serum was completely removed. Each tube having clot was again weighed to determine the clot weight. Each microcentrifuge tube containing clot was properly labeled and 100µl of clavix along with various dilutions of aqueous plant extracts 10 per cent, 20 per cent were added to the tubes. Water was also added to one of the tubes containing clot and this serves as a negative thrombolytic control. All the tubes were incubated at 37°C for 90 minutes. All the tubes were again weighed to observe the difference in weight after clot disruption. Differnce in weight taken before and after clot lysis was expressed as percentage of clot lysis (Prasad *et al.*, 2007).

2.6. Bio-safety Screening of the Plant Materials

Brine Shrimp Lethality Assay (BSLA) was carried out to investigate the cytotoxicity of extracts of fresh and dry *Aloe vera*. Brine shrimp (*Artemia salina*) was obtained by hatching brine shrimp eggs in artificial sea water (3.8 per cent non ionized sodium chloride solution) for 48 hours (Ved *et al.*, 2010).

3. Results and Discussion

3.1. Determination of Clotlysis in Fresh and Dry *Aloe vera*

Clot lysis of aqueous extract of fresh and dry *Aloe vera* gel was determined and the results are represented in Table 23.1.

Table 23.1: Thrombolytic activity of aqueous extract of fresh and dry *Aloe vera*.

Concentration of the Extract(mg/ml)	Thrombolytic Activity (per cent)	
	Aloe vera Fresh	Aloe vera Dry
Clavix	58.38±4.89	58.38±4.89
Water	10.75±7.86	10.75±7.86
10	18.18±8.20	15.06±7.85
20	20.79±9.88	19.57±9.24
CD (p<0.05)		8.17

Values are mean±S.D (n=20).

From the the table it is indicated that there was a significant increase in the thrombolytic activity of both fresh and dry *Aloe vera*, as the concentration of the sample increased.

Both fresh and dry *Aloe vera* gel showed lesser activity compared to positive control and better activity compared to negative control.

This was supported by Mannan *et al.* (2011) who reported that *Cassia alata* seed oil extract has moderate thrombolytic activity compared to negative control. Prasad *et al.* (2007) stated that the percent clot lysis of the *Fagonia arabica* plant extract was siginificantly higher when compared to the negative control (water).

3.2. Brine Shrimp Lethality Assay

Biosafety screening was carried out in fresh and dry *Aloe vera* plant samples and the results are represented in Tables 23.2 and 23.3.

The percentage lethality of brine shrimp increases with increase in concentration of the plant extracts. Dry *Aloe vera* exhibited 70 per cent lethality at the highest concentration (25 mg/ml) and 30 per cent at the lowest concentration (5 mg/ml) where as fresh *Aloe vera* exhibited 45 per cent and 10 per cent respectively. The findings revealed that both the extracts showed mild cytotoxicity against brine shrimp nauplii. Dry *Aloe vera* was found to be more Cytotoxic than the fresh *Aloe vera* extract.

Ved *et al.* (2010) mentioned that BSLA is an excellent choice for preliminary assessment of toxicity of herbal drugs. The findings of the present study are in agreement with Kamba *et al.*, 2010 who have indicated that the degree of lethality was found to be directly proportional to the concentration of the extract.

Table 23.2: Cytotoxic effect of fresh *Aloe vera* on Brine shrimp nauplii.

Sl.No.	Conc. (mg/ml)	No.of Shrimp Introduced	No.of Shrimp Survived after 24 hrs	Lethality (per cent)	Log10 Conc.	Probit Value	LC50
1.	Control	20	20	0	–	–	31.62
2.	5	20	18	10	0.6990	3.71	
3.	10	20	17	15	1.0000	3.96	
4.	15	20	15	25	1.1761	4.33	
5.	20	20	14	30	1.3010	4.48	
6.	25	20	11	45	1.3979	4.87	
7.	30	20	10	50	1.4771	5.00	

Table 23.3: Cytotoxic effect on dry *Aloe vera* on Brine shrimp nauplii.

Sl.No.	Conc. (mg/ml)	No.of Shrimp Introduced	No.of Shrimp Survived after 24 hrs	Lethality (per cent)	Log10 Conc.	Probit Value	LC50
1.	Control	20	20	0	–	–	15.14
2.	5	20	14	30	0.6990	4.48	
3.	10	20	12	40	1.0000	4.75	
4.	15	20	11	45	1.1761	4.87	
5.	20	20	8	60	1.3010	5.25	
6.	25	20	6	70	1.3979	5.52	

4. Conclusion

A number of medicinal plants are used in traditional system of medicine for the treatment of thrombosis and clot related diseases. Medicinal plant which is a priceless gift to human has been used all over the world for the treatment and prevention of various ailments. Hence a great emphasis has been laid on such valuable plant sources in lysing the blood clots which pose a great threat to mankind.

The present study has been designed to assess the thrombolytic and cytotoxic property of fresh and dry *Aloe vera* gel. Thrombolytic activity of the plant extracts increased with increase in concentrations. It was also seen that the aqueous extract of fresh *Aloe vera* showed greater thrombolytic activity compared to dry *Aloe vera*. Bio safety screening by BSLA indicated that the fresh *Aloe vera* was less Cytotoxic than the dry *Aloe vera*.

Hence the results suggest that the fresh and dry *Aloe vera* have potent thrombolytic potential and thus can be recommended for further analysis for the study of its effect on cardiovascular and other thrombus related diseases.

References

Anwar, S., Khan, I.N., Sarkar, M.I., Barwa, S., Kamal, M.A.T.M. and Hosen, Z.S.M., 2011. Thrombolytic and cytotoxic effect of different herbal extracts. *International Journal of Pharmaceutical Sciences and Research,* 2(12): 3118–3121.

Kamba, A.S. and Hassan, L.G., 2010. Antibacterial screening and brine shrimp toxicity of *Securidaca longepedunculata* root bark. *African Journal of Pharmaceutical Sciences and Pharmacy,* 1(1): 85–95.

Mannan, A., Kawser, J., Ahmed, A.M., Islam, N.N., Alam, M.S.M., Emon, M.A.E. and Gupta, S.D., 2011. Assessment of antibacterial, sthrombolytic and cytotoxic potential of *Cassia alata* seed oil. *Journal of Applied Pharmaceutical Science,* 1(9): 56–59.

Mosa, R.A., Lazarus, G.G., Gwala, P.E., Oyedeji, A.O. and Opoku, A.R., 2011. *In vitro* anti-platelet aggregation, antioxidant and cytotoxic activity of extracts of some zulu medicinal plant. *Journal of Natural Products,* (4): 136–146.

Prasad, S., Kashyap, R.S., Deopujari, J.Y., Purohit, H.J., Taori, G.M. and Daginawala, H.F., 2007. Effect of *Fagonia arabica* (Dhamasa) on *in vitro* thrombolysis. *B.M.C.,* 7: 36.

Shemy, H.A., Soud, M.A.M., Allah, A.A., Enein, K.M., Kabash, A. and Yogi, A., 2010. Antitumor properties and modulation of antioxidant enzymes activity by *Aloe vera* leaf active principles isolated via supercritical carbon dioxide extraction. *Current Medicinal Chemistry,* 17: 129–138.

Ved, C.H., More, N.S., Bharate, S.S. and Bharate, S.B., 2010. Cytotoxicity screening of selected Indian medicinal plants using brine-shrimp lethality bioassay. *Advances in Natural and Applied Sciences,* 4(3): 389–395.

Scientific Basis of Herbal Medicine (2013)
Editor: **Dr. Parimelazhagan Thangaraj**
Published by: **DAYA PUBLISHING HOUSE, NEW DELHI**

Pages **207–215**

Chapter 24

Phytochemical and Antimicrobial Activity of *Rotula aquatica,* Lour.

S. Sumithra, V. Sasikala and B. Vijayakumari

Department of Botany,
Avinashilingam University, Coimbatore – 641 043, Tamil Nadu, India

1. Introduction

The use of medicinal plants is growing worldwide because of the increasing toxicity and allergic manifestations of the synthetic drugs. *Rotula aquatica* Lour. belonging to the family Boraginaceae is represented by about 100 genera and 2000 species. The plant is scattered throughout peninsular and western ghats of India in the sandy and rocky beds of streams and rivers. The plant is a mandatory component of many ayurvedic drug preparations and is an important traditional medicine for kidney and bladder stones. The root and tuber are astringent, bitter and also useful in treating coughs, heart diseases, dysuria, blood disorders, fever, poisoning, ulcers and uterine diseases. Root decoctions are both diuretic and laxative and are used to treat bladder stones and sexually transmitted diseases.

The plant is reported to contain baunerol, steroid, alkaloid (Singh *et al.,* 2011). A decoction of roots of *R. aquatica* showed antiliathiatic activity in glycolic acid induced nephrolithiasis rats and the results were summarized based on the ionic change in both urine and serum (Christina *et al.,* 2002). Antidiabetic activity of the aqueous extract of *R. aquatica* was shown in alloxan induced diabetic rats (Pari *et al.,* 2002). In ayurveda, *R. aquatica* has been reported to be used for diabetes (Christina *et al.,* 2002), cardiotonic activity (Pankaj Oudhia, 2001), anti-urolithiatic activity (Reddy and Srinivasan, 2000). The aqueous extract of the root of *R. aquatica* Lour. showed antioxidant activity. It also contains sterol and rhabdiol (Khare, 2007)

A large number of research works on the phytochemistry, pharmacology and several other aspects have been conducted, but there have been no report on phytochemical screening and *in vitro* bioactivities of *R. aquatica.* So the present investigations were carried out to study the phytoconstituents and antimicrobial activities of petroleum ether, methanol, chloroform and aqueous extracts of R. *aquatica*.

2. Materials and Methods

2.1. Plant Collection and Identification

The fresh plant parts of R. *aquatica* Lour. were collected from Kuttiyadi (Malapuram district) in Kerala state. The collected plant material was identified and their authenticity was confirmed by comparing the voucher specimen at the herbarium of Botanical Survey of India, Southern circle, Coimbatore, Tamil Nadu, India. Freshly collected plant materials were cleaned to remove adhering dust and then dried under shade. The dried sample were powdered and used for further studies.

2.2. Preparation of Extracts

The air dried, powdered plant material was extracted with petroleum ether, methanol and aqueous using soxhlet apparatus. Each time before extracting with the next solvent, the material was dried in hot air oven below 40°C. Finally, the material was macerated using hot water with occasional stirring for 24 hr. The different solvent was evaporated using a rotary vacuum-evaporator (Yamato RE300, Japan) at 50°C and the remaining water was removed by lyophilisation (VirTis Benchtop K, USA).

2.3. Phytochemical Screening

The freshly prepared crude extract of R. *aquatica* were qualitatively tested for the presence of Alkaloids (Hager's test), Flavanoids (Ammonia test), Steroids (Salkowski test), Terpenoid (Salkowski test), Reducing sugars (Fehling's test), Saponins (Frothing test), Tannins ($FeCl_2$ test), Cardiac glycosides (Killer-Killani's test) and Anthraquinones (Chloroform layer test) (Harborne, 1973)

2.4. Biochemical Parameters

1. Estimation of protein (Lowry *et al.*, 1951)
2. Estimation of carbohydrates (Hedge and Hofreiter, 1962)

2.5. DPPH Radical Scavenging Activity

The free radical scavenging activity of R. *aquatica* extracts were measured by decrease in the absorbance of methanol solution of DPPH (2.2 Diphenyl – 1 – picrylhydrazyl) (Feresin *et al.*, 2002). A methonal solution of the sample extracts at various concentrations (100-500 µg) was added to 5 ml of 0.1 mm methanolic solution of DPPH and allows standing for 20 min at 27° C. The absorbance of the sample was measured at 570 nm. Radical scavenging activity was expressed as the inhibition percentage of free radical by the sample and was calculated using the formula.

% DPPH radical scavenging activity =
(control OD – sample OD/Control OD) × 100

2.6. Antimicrobial Assay

The antibacterial activity was carried out by the agar-well diffusion method (Smania *et al.*, 1995). Petroleum ether, chloroform and methanol extracts were the samples taken for antibacterial studies. Each bacterial suspension was spread over the surface of the nutrient agar plates by means of cotton swab containing 4 wells with 25 μl each of 25 percent concentration of the extract using micropipette. The plates were incubated at 37°C for 24 hours. The antimicrobial activity of the test agents were determined by measuring the diameter of zone of inhibition expressed in mm.

2.7. Statistical Analysis

Values were expressed as mean of triplicate analysis of the samples (n=3) standard deviation(SD). Analysis of variance and significant differences (P≤0.05) among means were tested by one-way ANOVA and Dunnet multiple range test.

3. Results and Discussion

3.1. Phytochemical Screening

Preliminary phytochemical screening showed (Table 24.1) the presence or absence of alkaloids, flavanoids, steroids, terpenoids, reducing sugars, saponins, tannins, cardiac glycosides, anthraqunones in varying amount in the *R. aquatica* extracts.

Table 24.1: Estimation of protein and carbohydrates content of root, stem and leaf.

Sl.No.	Samples	Protein (mg/100 g)	Carbohydrates (mg/100 g)
1.	Root	6.78±0.05	18.11±0.26
2.	Stem	5.72±0.09	12.90±0.07
3.	Leaf	3.52±0.08	15.72±0.32

Values are represented as mean±SD with triplicate estimation.

Previous studies reported the presence of alkaloids, tannins, carbohydrates, saponins, glycosides, proteins and aminoacids, phenolic compounds, flavonoids and terpenoids in selected ethno medicinal plants of Dindugul district (Karuppusamy and Karmegam, 2005). Presence of varieties of chemical compounds the significant amount of biological activities of *R. aquatica* extracts.

3.2. Estimation of Protein

The protein content of the *R. aquatica* was presented in (Table 24.2). Among the three samples, the root showed the highest amount of protein content (6.78 mg/100 g) followed by the leaf and stem.

3.3. DPPH Radical Scavenging Activity

From the analysis of Figure 24.1. It can be concluded that the scavenging effect of *R. aquatica* sample increases as the concentration increases. The root sample of *R. aquatica* showed (Figure 24.1) highest radical scavenging capacity followed by leaf and stem samples.

Based on the scavenging capacity of the free radicals (DPPH) the highest antioxidant activity was found in root sample. This could be due to the presence of

Table 24.2: Phytochemical compositions of *Rotula aquatica* extracts.

Phytoconstituents	Name of the Test	Leaf			Stem			Root		
		Petroleum Ether	Chloro-form	Methanol	Petroleum Ether	Chloro-form	Methanol	Petroleum Ether	Chloro-form	Methanol
Alkaloids	Hager's test	–	++	+	–	+	++	+	++	+++
Anthragunones	Chloroform layer test	+	–	++	+	++	++	+	++	++
Cardiac glycosides	Killer-Killani's test	+	++	+	–	+	+	++	+	++
Flavanoids	Ammonia test	+	++	++	+	+	++	+	++	++
Reducing sugars	Fehling's test	–	+	++	+	–	+	–	+	++
Saponins	Frothing test	+	++	+	+	++	+	+	++	++
Steroids	Salkowski test	+	+	+	+	–	+	+	+	++
Tannins	Fecl$_2$ test	+	+	++	–	+	+	–	++	+
Terpenoids	Salkowski test	–	+	+	+	–	++	–	+	++

+++: Highly present, ++: Moderately present, +: Slightly present, –: Absent.

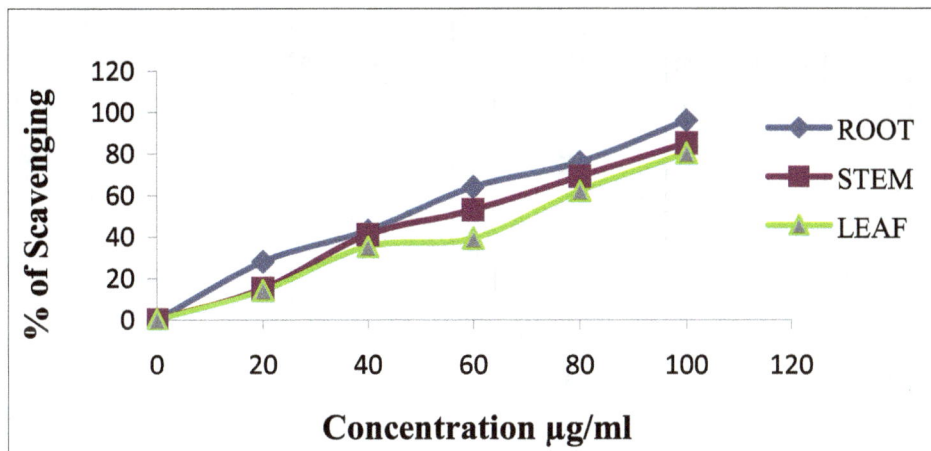

Figure 24.1: DPPH scavenging.

rich in polyphenols (tannins), which could be the possible explanation offered for its antioxidant activity (Patil *et al.*, 2004).

Table 24.3: Estimation of antioxidant activity using in *in-vitro* DPPH assay.

Sl.No.	Samples	Standard Value (Ascorbic acid)	Graded Concentrations of the Sample (µg/ml)				
			20	40	60	80	100
1.	Root	92.29±3.69	28.45±3.76	43.55±3.16	64.85±2.15	76.46±1.60	96.19±2.83
2.	Stem	92.29±3.69	15.83±2.57	41.70±3.21	53.05±1.48	69.27±1.34	85.63±3.69
3.	Leaf	92.29±3.69	14.53±1.98	35.14±1.95	39.84±1.49	62.79±1.64	80.01±2.28

Values are expressed by mean±SD of three replicates.

3.4. Antimicrobial Assay

Antimicrobial activities of the *R. aquatica* extracts were tested against three pathogenic organisms and the results are presented in Table 24.4. The different extracts of root (petroleum ether, chloroform and methanol) were studied by agar well diffusion and disc diffusion method. The highest inhibition zone of 0.7 mm in petroleum ether extract was observed against *Salmonella sp.* compared to the control (Chloramphenicol) 0.8 mm. The least inhibition zone was expressed against *K. pneumoniae* in petroleum ether extract of root (0.1 mm) by agar well diffusion method. In the disc diffussion method, maximum inhibition zone of 0.8mm was recorded in methanolic extract of root against *S. sonnei*. The inhibition zone registered against *Salmonella* sp. and *K. pneumoniae* were on par with each other (0.4 mm and 0.4 mm) in methanolic extract. The minimal inhibition zone of 0.1 mm was observed in petroleum ether extract against *Klebsiella* sp.

The extracts of stem (petroleum ether, chloroform and methanol) were studied by agar well diffusion method and disc diffusion method. The highest inhibition

Table 24.4: Inhibition zone by the extraction of root against bacteria (Agar well diffusion method and disc diffusion method)

Test Oragnisms	Agar Well Diffusion Method				Disc Diffusion Method			
	Zone Inhibition Method (mm)				Zone Inhibition Method (mm)			
	Petroleum ether	Chloroform	Methanol	Control	Petroleum ether	Chloroform	Methanol	Control
Klebsiella pneumonia	0.1	0.3	0.4	0.6	0.1	0.2	0.4	0.9
Salmonella sp.	0.7	0.4	0.6	0.8	0.2	0.4	0.4	0.8
Shigella sonnei	0.4	0.2	0.7	0.9	0.4	0.6	0.8	0.9

Control – Chloramphenicol.

Table 24.5: Inhibition zone by the extraction of stem against bacteria (Agar well diffusion method and disc diffusion method)

Test Oragnisms	Agar Well Diffusion Method				Disc Diffusion Method			
	Zone Inhibition Method (mm)				Zone Inhibition Method (mm)			
	Petroleum ether	Chloroform	Methanol	Control	Petroleum ether	Chloroform	Methanol	Control
Klebsiella pneumonia	0.3	0.9	0.7	1.0	0.2	0.7	0.4	1.0
Salmonella sp.	0.8	0.4	0.6	1.1	0.3	0.4	0.5	0.8
Shigella sonnei	0.3	0.7	0.5	0.9	0.3	0.4	0.3	0.6

Control – Chloramphenicol.

Table 24.6: Inhibition zone by the extraction of leaf against bacteria (Agar well diffusion method and disc diffusion method)

Test Oragnisms	Agar Well Diffusion Method				Disc Diffusion Method			
	Zone Inhibition Method (mm)				Zone Inhibition Method (mm)			
	Petroleum ether	Chloroform	Methanol	Control	Petroleum ether	Chloroform	Methanol	Control
Klebsiella pneumonia	0.3	0.7	0.5	1.0	0.1	0.2	0.2	0.4
Salmonella sp.	0.8	0.5	0.6	0.9	0.3	0.4	0.3	0.9
Shigella sonnei	0.4	0.5	0.7	0.9	0.3	0.5	0.7	0.9

Control – Chloramphenicol.

zone of 0.9 mm was exhibited against *K. pneumoniae* in chloroform extract and 0.8 mm in petroleum ether extract against *Salmonella* sp. compared to the control (Chloramphenicol) 1.1 mm. The least inhibition zone was expressed against *S. sonnei* in petroleum ether extract of stem (0.3 mm) in agar well diffusion method. In the disc diffussion method, maximum inhibition zone of 0.5mm was recorded in methanolic extract of stem against *Salmonella* sp. The minimal inhibition zone of 0.2 mm was observed in petroleum ether extract against *Klebsiella sp.* (Table 24.5).

Among the different extracts of leaf (petroleum ether, chloroform and methanol) by agar well diffusion method and disc diffusion method the highest inhibition zone of 0.8 mm was exhibited against *Salmonella* sp in petroleum ether extract and 0.4 mm in petroleum ether extract against *S. sonnei* compared to the control (Chloramphenicol) 0.9 mm The least inhibition zone was expressed against *K. pneumoniae* in petroleum ether extract (0.3 mm). In the disc diffussion method, maximum inhibition zone of 0.7 mm was recorded in methanolic extract of leaf against *S. sonnei*. The minimal inhibition zone of 0.1 mm was observed in petroleum ether extract against *Klebsiella* sp. (Table 24.6).

The antimicrobial study indicated that the crude extacts of *R. aquatica* had better antibacterial activities at higher concentrations against the tested microorganism. This may be due to the sufficient concentration of antimicrobial constituents in the solvent extracts. The antibacterial activity of different extracts from the leaves of *Ocimum gratissium* tested against *Staphylococcus aureus*, *E.coli*, *Salmonella typhi* and *S. typhimurium*, showed inhibitory effects on the selected bacteria and the minimum inhibitory concentration (MIC) ranged from 0.1 per cent for *S.aureus* to 0.01 per cent for *E. coli* and *S. typhimurium* and 0.001 per cent for *S. typhi* (Adebolu and Oladimeji, 2005).

4. Conclusion

The organic solvent extracts of *R. aquatica* is a very good source of phytochemicals and showed marked *in vivo* bioactivities which can offer remedies to some of the common ailments ranging from common cold to complex pathological disorders. Studies are in progress in the laboratory to unveil the active components responsible for the bioactivities of the plant with high medicinal value.

References

Adebolu, T.T. and Oladimeji Salau Abiola, 2005. Antimicrobial activity of leaf extracts of *Ocimum gratissimum* on selected diarrhoea causing bacteria in south western Nigeria. *African Journal of Biotechnology*, 4(7): 682–684.

Feresin, G.E., Tapia, A., Gutierrez, R, A., Delporte, C., Backhouse, E.N. and Schmeda-Hirschmann, G., 2002. Free radical scavengers, anti-inflammatory and analgesic activity of *Acaena magellanica*. *Journal of Pharmacy and Pharmacology*, 54(6): 835–844.

Harborne, J.B., 1973. *Phytochemical Methods: A Guide to Modern Techniques of Plant Analysis*. Chapman and Hall, New York, p. 279.

Hedge, J.E. and Hofreiter, B.T., 1962. *Determination of Total Carbohydrate Chemistry.* Academic Press, New York, p. 17.

Karuppusamy, S. and Karmegam, N., 2005. Screening of ethanomedicinal plants of Dindigul district (South India) for antimicribial activity. *Journal of Ecobiology,* 17(1): 455–459.

Khare, C.P., 2007. *Indian Medicinal Plant.* Springer International Edition, New Delhi.

Lowry, O.H., Rosenbourgh, N.J., Farr, A.L. and Randall, R.F., 1951. Protein measurement with folin-phenol reagent, *Journal of Biological Chemistry,* (7): 257–275.

Pankaj Oudhia, 2007. *Medicinal Herbs of Chattisgarh, India having Less Known Traditional Uses.* Articles online Rassivan, p. 135.

Pari, L.I. and Saravanan, G., 2002. Antidiabetic effect of congent DB, a herbal drug in alloxan induced diabetes mellitus. *Compar. Biochem. Physiol.,* 131(1): 19–25.

Patil, S., Jolly, C.I. and Narayanan, S., 2004. Evaluation of antimitotic activity of the root of *Rotula aquatica* (lour): A traditional herb used in the treatment of cancer. *Indian Journal of Experimental Biology,* 42: 893–899.

Reddy, G.B.S. and Sriniwasan, K.K., 2000. An experimental evaluation of root of *Rotula aquatica* for anti urolithiatic activity in albino rats. *Indian Drugs,* 30(8): 398–404.

Singh, S., Rai, A.K., Sharma, P. and Barshiliya, Y., 2011. Comparative study of anthelmintic activity between aqueous extract of *Aerva lanata* and *Rotula aquatica* Lour. *Asian Journal of Pharmacy and Life Science,* 1(3): 211–221.

Smania, D.L. and Salunkhe, D.K., 1995. Free Radicals and antioxidants in food and in vivo. *Journal of Food Science and Nutrition,* 35(4): 7–20.

Scientific Basis of Herbal Medicine (2013)
Editor: Dr. Parimelazhagan Thangaraj
Published by: DAYA PUBLISHING HOUSE, NEW DELHI

Pages 217–227

Chapter 25

Therapeutic and Medicinal Value of Vegetables: A Review

C. Thangamani, C. Kavitha and L. Pugalendhi

Horticultural College and Research Institute,
Tamil Nadu Agricultural University, Coimbatore – 641 003, India

1. Introduction

India is endowed with variety of vegetables. Vegetables are the cheap source of proteins, vitamins, minerals and essential amino acids. They are included in meals mainly for their nutritional value, however some are reserved for the sick and convalescence because of their medicinal properties. "The art of medicine consists of entertaining the patient while, Nature cures the disease". The medicine of the future will no longer be remedial, it will be preventive, not based on drugs but on the best diet for health. If the body has the ability to heal itself, it will use the raw materials found in foods to do its own healing work. Herbs do not heal, they feed. Herbs do not force the body to maintain and they feed the body and repair itself, and simply support the body in those natural functions.

The World Health Organization acknowledges that the global intake of vegetables is less than 20-50 per cent of the recommended amount. In developed countries, the significantly low vegetable intake is due to the consumer's preferences for convenience foods and not the scarcity of the vegetables. However, in the US more processed vegetables are consumed than fresh vegetables (Rickman et al., 2007). A high intake of food rich in natural antioxidants has been shown to increase the antioxidant capacity of the plasma and reduce the risk of some, but not all, cancers, heart diseases and stroke. Recent research has also shown that through overlapping or complementary effects, the complex mixture of phytochemical compounds in fruits and vegetables provides a protective effect on health than single phytochemicals.

1.1. Vitamins

These substances are essential for normal body functions, cell function regulation, growth and development. They must be obtained from the diet, as the body cannot produce them in adequate amounts. Vitamins such as A, B6, C, and K are important to the human body and can be provided by vegetables.

1.2. Antioxidants

Antioxidants protect against highly reactive metabolic byproducts (known as free radicals) that cause cell damage in the human body. Vitamins, minerals, and phytochemicals contained in fruits and vegetables each have antioxidant activity. Carotenoids, selenium, Vitamins C and E are all examples of antioxidants. They have been investigated for their specific role in the prevention of cancers, heart disease, eye disease and other human health conditions.

1.3. Phytochemicals

Plant chemicals considered to be beneficial to health but are not essential nutrients are called phytochemicals. Carotenoids and flavonoids are examples of compounds that are considered phytochemicals. Associations between disease prevention and individual phytochemicals remain unproven, although many studies show the benefit of a diet high in fruits and vegetables. Further research is needed to directly attribute specific health benefits to specific compounds.

1.4. Carotenoids

Carotenoids are a class of pigments that are responsible for giving plants a red, yellow or orange colour. Lutein, beta-carotene, lycopene and zeaxanthin are examples of carotenoids that are important to the human diet. For optimal absorption in the human body, they are best consumed cooked with a little fat after they have been chopped or pureed. They are a very important source of vitamin A. They are being investigated for their role in heart disease, some cancers, and eye disease.

1.4.1. Lycopene

Lycopene gives tomatoes and some fruits their red colour. Most lycopene in our diet is obtained from cooked and processed tomato products. Lycopene may play a preventative role in certain cancers and heart disease.

1.4.2. Lutein and Zeaxanthin

Dark green and leafy vegetables are the predominant source of lutein and zeaxanthin in the human diet. They may play a role in preventing oxidative damage to the eye, and may reduce the risk of age related macular degeneration.

1.4.3. Beta-Carotene

Beta-carotene is responsible for the orange and yellow colour often seen in fruits and vegetables. Carrots, squash, sweet potatoes and spinach are good sources of beta-carotene. It has been investigated for its role in the prevention of cardiovascular disease and certain cancers.

1.5. Flavonoids

This is a large family of phytochemical compounds produced by plants. High intake of flavonoid rich foods have been shown to reduce the risk of cardiovascular disease, but whether this is due to specific, individual compounds remains to be proven. Anthocyanins are a subclass of flavonoids that give red, blue and purple berries and grapes their colour. They have been associated with improving blood vessel health in humans. Quercetin is one of the most widely distributed flavonoids in the human diet, found in apples, onions and citrus fruits. It may have antioxidant and anti-inflammatory activity, but firm conclusions on its role in heart disease, arthritis and eye disease still need to be shown.

This article reviews the medicinal properties of important vegetables in the groups *viz.*, cucurbitaceous, cruciferous, some leafy vegetables, solanaceous and alliaceae with relevant reports. Since these vegetable are consumed habitually by all the people in their regular diets.

2. Review of Literature and Discussion

2.1. Cucurbitaceous Vegetables

Cucurbitaceous vegetables include bitter gourd, snake gourd, cucumber, ridge gourd, melons and pumpkins etc. In recent years, the importance of Cucurbitaceous species, a commonly used vegetable has been recognized in the experimental control of diabetes mellitus. A wide range of plant-derived principles belongs to compounds, mainly alkaloids, glycosides, galactomannan gum, polysaccharides, hypoglycans, peptidoglycans, guanidine, steroids, glycopeptides and terpenoids. However, many other active agents obtained from plants still needs to be well characterized.

2.1.1. Bitter Gourd (*Momordica charantia* L.)

Bitter gourd is very commonly used cucurbits and possesses various well documented medicinal potentials in its different parts. It has been used in the ancient traditional medicine of China, India, Africa and Latin America. Bitter gourd extracts possess anti oxidant, anti microbial, antiviral and anti-ulcerogenic properties and also having the ability to lower blood sugar (Raman and Lau, 1996). A mixture of steroidal saponins knows as charatin (Insulin like peptides), as well as alkaloids, appears to be responsible for the hypoglycemic action in bitter gourd extracts. Hypoglycemic effects of these chemicals are more pronounced in fruit, where they are present in abundance.

Vast researches have claimed the presence of different active principles classified into different chemical types responsible for its famous antidiabetic and anti hyperlipidemic characters. The bio active compounds from bitter gourd activate a protein called AMPK, which is well known for regulating fuel metabolism and enabling glucose uptake, processes which are impaired in diabetics. Other chemical constituents in bitter gourd are furnished below.

Table 25.1: Chemicals constituents of bitter gourd.

Source	Phytochemicals	Reference
Plant body	Momorcharins, momordenol, momordicilin, momordicins, momordicinin, momordin, momordolol, charantin, charine, cryptoxanthin, cucurbitins, cucurbitacins, cucurbitanes, cycloartenols, diosgenin, elaeostearic acids, erythrodiol, galacturonic acids, gentisic acid, goyaglycosides, goyasaponins, multiflorenol.Glycosides, saponins, alkaloids, fixed oils, triterpenes, proteins and steroids	Husain *et al.,* 1994, Xie *et al.,*1998, Yuan *et al.,* 1999, Parkash *et al.,* 2002, Murakami *et al.,* 2001, Raman and Lau, 1996
Fruit	Momordicine, charantin, polypeptide- p insulin, ascorbigen, Amino acids – aspartic acid, serine, glutamic acid, threonine, glutamic acid, threonine, alanine, g-amino butyric acid and pipecolic acid, luteolin, Fatty acids – Lauric, myristic, palmitic, palmitoleic, stearic, oleic, linoleic, linolenic acid	Lolitkar and Rao, 1998, Yuwai *et al.,* 1991

Case Study on Anti-diabetic Property of Bitter Gourd

Aqueous extract of the leaves of *Momordica charantia* was tested for hypoglycemic activity by oral administration to both normal rats and alloxan-induced diabetic rats for 28 days. Wistar rats were divided into 4 groups A, B, C, D. Group A (control) was treated with intraperitoneal injection of the vehicle (saline alone), group B alloxan-induced diabetic rats administered orally equal volumes of vehicle (distilled water) alone, group C, normal rats and group D, alloxan-induced diabetic rats both administered 400 mg/kg of the aqueous extract. On the 29th day, blood glucose, plasma insulin and the effect on Oral Glucose Tolerance Test (OGTT) were monitored at 3rd, 6th and 9th hour, for 50 per cent of the rats of each group, while the rest of the rats were sacrificed and the levels of certain biochemical parameters in the serum and examination of some visceral organs were investigated. 'At the end of the experiment, the blood glucose level of group C rats was found to have reduced significantly ($p<0.01$) while there was no significant difference in the blood glucose level of group D rats. Significant reduction in blood glucose during OGTT was observed in group D rats at 3rd, 6th and 9th hour, compared to non treated diabetic rats. No changes were observed in plasma insulin levels, indicating that the probable hypoglycemic mechanism involves improved glucose tolerance by preventing glucose from being reabsorbed into the intestines. Highly significant increase in alkaline phosphatase ($p<0.001$), AST and ALT ($p<0.001$) in alloxan-induced diabetic rats (B and D) and in normal rats administered the aqueous extract, was observed, indicating liver hyperactivity. Significant increases in creatinine clearance and a significant decrease in creatinine ($p<0.01$) in group C suggest proper functioning of the kidneys. Macroscopic examination of some visceral organs revealed profound pathological differences in diabetic rats and normal rats administered the aqueous extract (Protus *et al.,* 2012).

2.1.2. Bottle Gourd (*Lagenaria siceraeria* M.)

The entire plant is recognized to be beneficial in ethnic systems of medicine. The fruit is sweet, diuretic, antipyretic, antibilious, tonic for the liver, vulnerary, and antiperiodic. It can cure blood diseases in persons of pitta constitution; muscular pain and dry cough. In Punjab, the pulp is applied to the soles of the feet of those with "burning feet". The seeds are fattening, cooling, anthelmintic, and a brain tonic; they can cure cough, fever, scalding urine, and earache; they also reduce inflammation (Unani). Their oil can be applied for headache. The rind of the fruit is good for piles, while its ash is styptic and vulnerary. The root is applied in the treatment of dropsy (Kirtikar and Basu, 2005).

The fruits, leaves, oil, and seeds are edible and used by local people as folk medicines in the treatment of jaundice, diabetes, ulcer, piles, colitis, insanity, hypertension, congestive cardiac failure and skin diseases. The fruit pulp is used as an emetic, sedative, purgative, cooling, diuretic, antibilious and pectoral. The flowers are an antidote to poison. The stem bark and rind of the fruit are diuretic. The seed is vermifuge. Extracts of the plant have shown antibiotic activity. Leaf juice is widely used for baldness (Duke and Ayensu, 1985). Bottle gourd juice is an excellent remedy for heart problems, digestive and urinary disorders, and in diabetes. Dietary fiber present in bottle gourd helps in constipation, flatulence and even in piles. Topical application of a mixture of fruit juice and sesame oil on scalp gives beneficial results in baldness (hair loss). The juice also shows better effects in the treatment of insomnia, epilepsy, and other nervous diseases. Moreover it helps break up calculus (stones) in the body. In summer or hot conditions, bottle gourd fruit juice prevents excessive loss of sodium, satiating thirst, and giving a cooling effect (Rahman, 2003).

2.1.3. Ivy Gourd (*Coccinia indica*)

The aqueous and organic solvent (Petroleum ether, chloroform and ethanol) extracts from the leaves of *Coccinia indica* were tested against *Enterobacter aerogenes*, *Pseudomonas aeruginosa*, *Staphylococcus epidermidis*, *Bacillus subtilis* and *Salmonella typhimurium* by agar well diffusion method and broth dilution method. Results showed promising antibacterial activity against the bacteria tested. Among these, ethanol and aqueous extracts were found to have a more potent inhibitory effect comparing with the other extracts. Which prove the potentiality of the plant extracts for the treatment of various skin and gastrointestinal infections in humans. (Arshad *et al.*,2010).

2.2. Cruciferous Vegetables

Cruciferous vegetables includes cabbage, cauliflower, brussels sprout, sprouting broccoli, kale, turnip, radish and water cress etc. They rich in nutrients, including several carotenoids (beta-carotene, lutein, zeaxanthin); vitamins C, E, and K; folate; and minerals and also are a good fiber source. In addition, cruciferous vegetables contain a group of substances known as glucosinolates, which are sulfur-containing chemicals. These chemicals are responsible for the pungent aroma and bitter flavor of cruciferous vegetables. During food preparation, chewing, and digestion, the glucosinolates in cruciferous vegetables are broken down to form biologically active compounds such as indoles, nitriles, thiocyanates, and isothiocyanates. Indole-3-carbinol (an indole) and sulforaphane (an isothiocyanate) have been most frequently examined for their anticancer effects. Indoles and isothiocyanates have been found to inhibit the development of cancer in several organs in rats and mice, including the bladder, breast, colon, liver, lung, and stomach. Studies in animals and experiments with cells grown in the laboratory have identified several potential ways in which these compounds may help to prevent cancer.

A few studies have shown that the bioactive components of cruciferous vegetables can have beneficial effects on biomarkers of cancer-related processes in people. For example, one study found that indole-3-carbinol was more effective in reducing the growth of abnormal cells on the surface of the cervix. In addition, several case studies have shown that specific forms of the gene that encodes glutathione S-transferase, which is the enzyme that metabolizes and helps eliminate isothiocyanates from the body, may influence the association between cruciferous vegetable intake and human lung and colorectal cancer risk.

Table 25.2: Cruciferous vegetables and their medicinal uses.

Common Name	Botanical Name	Medicinal Uses and Health Benefits	Reference
Cabbage	*Brassica oleraceae* var. *capitata*	Skin diseases, anti microbial	Ogunlesi *et al.* (2010)
Cauliflower	*Brassica oleraceae* var. *bortistis*	Rich in folate and ascorbic acid	
Lettuce leaves	*Lactuca sativus*	Diuretic, anti-inflammatory	
Broccoli	*Brassica oleraceae* var. *italica*	High antioxidant activity, isothicyanate is anti-carcenogenic	Steck *et al.* (2007)
kale	*Brassica oleraceae* var. *acephala*	Rich in carotenoids, neutra-ceutical in kale reduce lung cancer and eye disorders	Kopsell *et al.* (2007)

2.3. Leafy Vegetables or Pot Herbs

Utilization of leafy vegetable is part of the food culture of the Indian house hold. Leafy vegetables have long been known and reported to have health protecting properties and uses. They have history of being consumed by the humans as both food and medicine. However, due to urbanization, the consumption of these vegetables appears to be declining.

The nutrient content of different types of vegetables varies considerably and they are not major sources of carbohydrates compared to the starchy foods which form the bulk of food eaten, but contain vitamins, essential amino acids, as well as minerals and antioxidants. Free radicals are generally unstable reactive molecules that are produced in animals and humans under physiological and pathological conditions. There is increased scientific evidence that oxidative stress which results in the generation of free radicals contributes to many common ailments including cancer, cardiovascular disease, cataract formation, as well as accelerating the ageing process. Epidemiological studies have shown a strong and consistent protective effect of dietary antioxidants against the risk of such illnesses. This protective effect is often attributed to different antioxidant components, such as vitamin C, vitamin E, carotenoids, polyphenolic compounds and other phytochemicals. A high intake of food rich in natural antioxidants has been shown to increase the antioxidant capacity of the plasma and reduce the risk of some cancers, heart diseases and stroke.

Table 25.3: List of some leafy vegetables and medicinal uses.

Common Name	Botanical Name	Medicinal Uses and Health Benefits	Reference
Amaranthus Green amaranth	*Amaranthus cruentus* L. *Amaranthus viridis* L.	Tapeworm expellant, relief of respiratory diseases, eye diseases and gonorrhea	Mensah *et al.,* 2008
Vallarai	*Centella asiatica* L.	Plant paste and juice are used in mental weakness and skin diseases	Bhagawati and Vandana Shiva, 2005
Dhaniya	*Coriandrum sativum* Linn.	Leaf and fruits are used as condiment and leaf paste is used in skin disease	Bhagawathi and Vandana Shiva, 2005
Black night shade/ Garden egg	*Solanum nigrum*	Leaf paste and branches are used in jaundice and high fever	Bhagawathi and Vandana Shiva, 2005
Indian spinach	*Basella rubra*	Fertility enhancement in women	Bhagawathi and Vandana Shiva, 2005
	Basella alba	Laxative	Ogunlesi *et al.,* 2010
Murungai	*Moringa oleifera* L.	Cures fever	Ogunlesi *et al.,* 2010
Roselle/ Pulichhakeerai	*Hibiscus sabdariffa*	Cough, diuretic and dressing of wounds	Ogunlesi *et al.,* 2010

2.4. Solanaceous Vegetables

2.4.1. Tomato (*Lycopersicon esculentum* M.)

All over the world, tomatoes occupy a significant position in agriculture and as well human diet. Second to potatoes, tomatoes are the most produced and consumed vegetable (Moreno *et al.,* 2006).Though tomatoes are eaten fresh it has been established that over 80 per cent of tomatoes are eaten in the form of the following by-products (Rao *et al.,* 1999), tomato juice, paste, puree, ketchup, sauce and soups.

Tomato is not only the largest source of lycopene, but also ranks third and fourth as the source of vitamin C and vitamin A respectively. Compared to other vegetables it has high levels of lycopene, folate, vitamin C and vitamin E. The carotenoids found in carrots and tomato were studied by a team of Harvard researchers. An analysis of over 1,24,000 men and women participating in a ten-year study found that those whose diets were high in lycopene and alpha-carotene, had a 20-25 per cent lower risk of lung cancer, especially among current smokers (Willoughby, 2000).

2.5. Alliaceae Vegetables

2.5.1. Onion (*Allium cepa* var. *cepa*)

Over 4000 years, Central Asian, Egyptian and Indian traditional systems of medicine have recognized the medicinal properties of onion, therefore it forms an inseparable part of the diet in these countries. Onions are rich in thiosulfinates, sulfides, sulfoxides, and other odoriferous sulfur compounds. The cysteine sulphoxides like allyl sulphoxide are responsible for the onion flavour and the eye-irritation that invariably occurs when one cuts an onion. The red colour of onion is due to anthocyanin while yellow due to quercetin. The carbohydrates found in onions are fructose, glucose and sucrose (Mogren *et al.*, 2007). The antioxidant flavonoids that are present in onions may reduce heart disease. The flavonol called quercetin is found in high levels in onions mostly in the outer dry scales. According to Griffiths *et al.* (2002), the onion has anti-carcinogenic, antithrombotic, antiplatelet, antiasthmatic and antibiotic abilities. The theosulfinates in onions have anti-microbial properties. Furthermore, the onion is effective against some bacteria like *Bacillus subtilis*, *Salmonella* and *E. coli*.

Onions are very low in calories (just 40 cal per 100 g) and fats; but rich in soluble dietary fiber. Onion phyto-chemical compounds *allium* and *Allyl disulphide* convert to allicin by enzymatic reaction when the bulb disturbed (crushing, cutting etc). Studies have shown that these compounds have anti-mutagenic (protects from cancers) and anti-diabetic properties (helps lower blood sugar levels in diabetics). Laboratory studies show that *allicin* reduces cholesterol production by inhibiting *HMG-CoA reductase* enzyme in the liver cells. Further, it also found to have anti-bacterial, anti-viral, and anti-fungal activities. *Allicin* also decreases blood vessel stiffness by release of nitric oxide (NO); thereby bring reduction in the total blood pressure. It also blocks platelet clot formation and has fibrinolytic action in the blood vessels which, helps decrease overall risk of coronary artery disease (CAD), peripheral vascular diseases (PVD), and stroke.

2.5.2. Garlic (*Allium cepa* var. *sativum*)

Allicin (thiosulfinate) is a compound found in garlic that contains sulfur and is known for its health benefits to humans. This compound gives garlic its characteristic taste and aroma. It is produced when the physical structure of the bulb is changed through, for example, crushing or cutting. Ali *et al.* (2000) stated in their report that there is some inconsistency regarding the effectiveness of garlic against hypertension. Conversely, a study by Mousa and Mousa (2007) stated that when garlic is taken together with vitamin C, it brings down marginally high blood pressure. Garlic may have the ability to retard the development of oral pathogens and so may be effective against periodontis (Bakri and Douglas, 2005). Garlic is said to have the ability to affect cardiovascular, cancer, hepatic and other microbial infections (Banerjee *et al.*, 2003).

Table 25.4: TCMD (Traditional Chinese Medicine Documentation) documented vegetables and functional components.

Compound	Activity	Source
Agavasaponin C	Platelet aggregation inhibitory	Garlic (*Allium sativum L.*)
Allicin	Antihypertensive; Antithrombotic	Shallot (*Allium fislulosum L.*); Garlic (*Allium sativum L.*)
Alliin	Antithrombotic; Platelet aggregation inhibitory	Onion (*Allium cepa L.*); Garlic (*Allium sativum L.*)
Bergapten	Antihypertensive	Tomato (*Lycopersicon esculentum* Miller)
Ferulic acid	Platelet aggregation inhibitory	Onion (*Allium cepa L.*)
Isoeruboside B	Platelet aggregation inhibitory	Garlic (*Allium sativum L.*)
Lycopene	Antiatherosclerotic	Tomato (*Lycopersicon esculentum Miller*); Bitter gourd (*Momordica charantia L.*)
p-Coumaric acid	Antilipemic	Potato (*Solanum tuberosum L.*)
Proto-iso-eruboside B	Antithrombotic	Garlic (*Allium sativum L.*)
Solasonine	Platelet aggregation inhibitory	Capsicum (*Capsicum annuum L.*); Eggplant (*Solanum melongena L.*)
Tomatine	Antihypertensive	Tomato (*Lycopersicon esculentum Miller*)
2-Vinyl-4H-1,3-dithiin	Platelet aggregation inhibitory; Antithrombotic; 5-lipoxygenase inhibitory	Garlic (*Allium sativum L.*)

3. Conclusion

This article reveals the important phytochemical contents and medicinal properties of routinely used vegetables in nutshell. Vegetables serve as a source for improving the health status as a result of the presence of various compounds vital for good health. However some active principles and their medicinal properties are yet to be explored. Hence phytochemistry studies on vegetables have a great scope in future not only as a research but also a science of practical utility for the good wellbeing of human being.

References

Ali, M., Al-Qattan, K.K., Al-Enezi. F., Khanafer, R.M.A. and Mustafa, T., 2000. Effect of allicin from garlic powder on serum lipids and blood pressure in rats fed with a high cholesterol diet. *Prostaglandins, Leukotrienes and Essential Fatty Acids.* 62: 253–259.

Arshad Hussain, Shadma Wahab, Zarin and Sarfaraj Hussain, M.D., 2010. Antibacterial Activity of the Leaves of *Coccinia indica* (W. and A) Wof India. *Adv. in Biological Research,* 4(5): 241: 248.

Bakri, I.M., and Douglas, C.W.I., 2005. Inhibitory effect of garlic extract on oral bacteria. *Archives of Oral Biology,* 50: 645–651.

Banerjee, S.K., Mukherjee, P.K. and Maulix, S.K., 2003. Garlic as an antioxidant: The good, the bad and the ugly. *Phytotherapy Research*. 17: 97–106.

Bhawati Uniyal and Vandana Shiva, 2005. Traditional Knowledge on medicinal plants among rural women of the Garhwal Himalaya, Uttanchal. *Indian Journal of Traditional Knowledge* 4(3): 259–266.

Duke, J.A. and Ayensu, E.S., 1985. *Medicinal Plants of China*, Vol. 2. Algonac, Michigan: Reference Publications.

Griffiths, G., Trueman, L., Crowther, T., Thomas, B. and Smith, B., 2002. Onions: A global benefit to health. *Phytotherapy Research*. 16: 603–615.

Husain, J., Tickle, I.J. and Wood, S.P., 1994. Crystal structure of momordin, a type I ribosome inactivating protein from the seeds of *Momordica charantia*. *FEBS Letters*, 342: 154–158.

Kirtikar, K.R., Basu, B.D., Dehradun, India: Oriental Enterprises, International Book distributors; 2005. *Indian Medicinal Plants*. pp. 1116–1167.

Kopsell, D.A., Kopsell, D.E. and Curran-Celentano, N., 2007. Carotenoid pigments in kale are influenced by nitrogen concentration and form. *Journal of Science Food and Agriculture*, 87: 900–907.

Lolitkar, M.M. and Rao, M.R.R., 1966. Pharmacology of a hypoglycaemic principle isolated from the fruits of *Momordica charantia*. Linn. *Indian Journal of Pharmacy*, 28: 129–133.

Mensha, J.K., Okoli, R.I., Ohaju Obodo, J.O. and Eifediyi, E., 2008. Phytochemical nutritional and medicinal prpoeprties of some leafy vegetables consumed by Edo people of Nigeria. *African Journal of Biotechnology*, 7(14): 2304–2309.

Mogren, L., 2006. Quercetin content in yellow onion (*Allium cepa* L.): Effects of cultivation methods, curing and storage. *Doctoral Dissertation*. <http://dissepsilon.slu.se/archive/00001246/01/Acta96.pdf> (24 October 2007).

Mousa, A.S., and Mousa, S.A., 2007. Cellular effects of garlic supplements and antioxidant vitamins in lowering marginally high blood pressure in humans. *Science Direct Nutrition Research*. 27: 119–123.

Murakami, T., Emoto, A., Matsuda, H. and Yoshikawa, M., 2001. Medicinal foodstuffs. Part XXI. Structures of new cucurbitane–type triterpene glycosides, goyaglycosides-a, -b, -c, -d, -e, -f, -g, and -h, and new oleanane–type triterpene saponins, goyasaponins I, II, and III, from the fresh fruit of Japanese *Momordica charantia* L. *Chemical and Pharmaceutical Bulletin (Tokyo)*, 49: 54–63.

Ogunlesi, M., Okiei, W., Azeez, L., Obakachi, V., Osunsami, M. and Nkenchor, G., 2010. Vitamin C contents of tropical vegetables and foods determined by voltammetric and titrimetric methods and their relevance to the medicinal uses of the plants. *International Journal of Electrochemical Science*, 5: 105–115.

Parkash, A., Ng, T.B. and Tso, W.W., 2002. Purification and characterization of charantin, a napin-like ribosome-inactivating peptide from bittergourd (*Momordica charantia*) seeds. *Journal of Peptide Research*, 59: 197–202.

Protus, A. T. and Ofogba, C. J., 2012. Evaluation of the Hypoglycemic Activity and Safety of *Momordica charantia* (Cucurbitaceae). *African Journal of Pharmaceutical Sciences and Pharmacy* 3(1): 17–29.

Rahman, A.S., 2003. Bottle Gourd (*Lagenaria siceraria*): A vegetable for good health. *Natural Product Research*, 2: 249–50.

Raman, A., and Lau, C., 1996. Anti-diabetic properties and phytochemistry of *Momordica charantia* L. (Cucurbitaceae). *Phytomedicine*, 2: 349–362.

Rao, A.W., Waseem, Z. and Agarwal, S., 1999. Lycopene content of tomatoes and tomato products and their contribution to dietary lycopene. *Food Research International*, 31: 737–741.

Rickman, J.C., Barrett, D.M. and Bruhn, C.M., 2007. Review: Nutritional comparison of fresh, frozen and canned fruits and vegetables. Part 1: Vitamin C and B and phenolics compounds. *Journal of Science, Food and Agriculture*, 87: 930–944.

Sanchez-Moreno, C., Cano, M.P. de Ancos, B., Plaza, L. Olmedilla, B. Granado, F. and Martin, A., 2006. Mediterranean vegetable soup consumption increases plasma vitamin C and decreases F2–isoprotanes, prostaglandin E2 and monocyte chemotactic protein-1 in healthy humans. *Journal of Nutritional Biochemistry*, 17: 183–189.

Steck, S.E., Gammon, M.D., Hebert, J.R., Wall, D.E. and Zeisel. S.H., 2007. *GSTM1, GSTT1, GSTP1,* and *GSTA1* Polymorphisms and Urinary Isothiocyanate Metabolites following Broccoli Consumption in Humans. *Journal of Nutrition,* 137: 904–909.

Willoughby, 2000. Tomatoes and carrots may reduce lung cancer risk. *American Vegetable Grower*, 48: 9.

Xie, H., Huang, S., Deng, H., Wu, Z., and Ji, A., 1998. Study on chemical components of *Momordica charantia,* 21: 458–459.

Yuan, Y.R., He, Y.N., Xiong, J.P., and Xia, Z.X., 1999. Three-dimensional structure of beta-momorcharin at 2.55 A resolution. *Acta Crystallographica Section DBiological Crystallography,* 55: 1144–1151.

Yuwai, K.E., Rao, K.S., Kaluwin, C., Jones, G.P. and Rivett, D.E., 1991. Chemical composition of *Momordica charantia* L. fruits. *Journal of Agricultural and Food Chemistry*, 39: 1762–1763.

Scientific Basis of Herbal Medicine (2013)   *Pages* **229–234**
Editor: Dr. Parimelazhagan Thangaraj
Published by: DAYA PUBLISHING HOUSE, NEW DELHI

Chapter 26

Analgesic Property of *Rubus ellipticus* Smith. Leaf Methanol Extract

Blassan P. George and Thangaraj Parimelazhagan

*Department of Botany, Bharathiar University,
Coimbatore – 641 046, Tamil Nadu, India*

1. Introduction

Natural products or drugs of natural origin are believed to be important source of new chemical substances which have potential therapeutic effects. Medicinal plants are one of the important sources, are extensively investigated both *in vitro* and *in vivo* to examine the potential activities. Most people living in the developing countries are almost completely dependent of the traditional medicinal practices for their primary health care needs and higher plants are known to be the main source for drug therapy in traditional medicine (Calixto, 2005). The genus *Rubus* is very diverse, includes over 750 species in 12 subgenera, and is found on all continents except Antarctica (Finn, 2008).

Due to the ethnomedicinal richness; *Rubus* species has been used in folk medicine (Patel *et al.*, 2004). *R. ellipticus* root paste is used as poultice for the treatment of bone fracture, applied on forehead during severe headache; fruit is edible (Pradhan and Badola, 2008). Ripe fruits are laxative and are used in the case of constipation, paste of young fruits are taken in case of gastritis, diarrhea and dysentery (Maity, 2004). The root juice drunk against urinary tract infection and its fruits are edible and were listed in the top ten wild edible medicinal plants in Tanahun district of Western Nepal (Uprety *et al.*, 2011). *R. ellipticus* is used for curing different ailments by the Lepcha tribe of Dzongu valley in North Sikkim, India. The young shoot is chewed raw to relieve sudden stomach pain. Root decoction given to the children to get rid of stomach warm. The inner root bark of the plant is valued as a medicinal herb in traditional Tibetan medicine, including its use as a renal tonic and antidiuretic (Pfoze, *et al.*, 2012).

Even though this plant has immense ethnomedicinal value; a survey of literature revealed that the analgesic activity of this plant using animal models has not yet been evaluated. Keeping this in view, the present investigation has been undertaken to study the central and peripheral analgesic activities of *R. ellipticus* leaf methanol extract to put forward a scope to develop an effective drug.

2. Materials and Methods

2.1. Plant Material Collection and Identification

The fresh plant parts of *R. ellipticus* were collected from Shola forest of Marayoor, Kerala, India, during the month of September 2010. The collected plant material was identified and authenticated by (Voucher specimen No. BSI/SRC/5/23/2010-11/Tech.1659) Botanical Survey of India, Southern circle, Coimbatore, Tamil Nadu.

2.2. Processing and Extraction

Collected plants were cleaned properly, separately shade dried and powdered. The powdered leaf was extracted in Soxhlet apparatus using methanol. The extract was concentrated to dryness under reduced pressure in a rotary evaporator to yield dried methanol extract.

2.3. Animals and Acute Toxicity Study

Healthy Swiss albino mice (25-30 g) of either sex and of approximately the same age were used for the study. They were fed with standard chow diet and water *ad libitum* and were housed in polypropylene cages in a well maintained and clean environment. The experimental protocol was subjected to scrutiny of institutional animal ethical committee for experimental clearance (KMCRET/Ph.D/03/2011).

The acute toxicity was performed as per Organization for Economic Co-operation and Development guidelines (OECD guidelines 423, 2001). The methanol extract at dose of 100, 500, 1000 and 2000 mg/kg was administered to mice in a single dose orally. Animals are observed individually after drug administration at least once during the first 30 minutes, periodically during the first 24 h, with special attention given during the first 4 h, and daily thereafter, for a total of 14 days.

2.4. Analgesic Activity

Peripheral and central analgesic activities of the tested extract were carried out in albino mice using acetic acid-induced writhing and hot plate methods, respectively.

2.4.1. Acetic Acid Induced Writhing Test

Male Swiss albino mice were divided into four groups with six animals each. Group 1 served as control, group 2 received standard drug Aspirin (100 mg/kg); group 3 and 4 received leaf methanol extract at doses of 200 mg/kg and 400 mg/ kg. The acetic acid 0.6 per cent v/v (10 ml/kg, i.p.) was injected intraperitoneally 1 h after administration of the drugs. After administration of acetic acid, number of writhes (abdominal muscle contractions) was counted over a period of 15 min and immediately after acetic acid injection (0 time) (Vogel and Vogel, 1997; Vyas *et al.*, 2008). Analgesic activity was expressed as the percentage protection against writhing produced by the tested extract compared with writhing at 0 time. Aspirin and methanol extract were suspended in carboxy methyl cellulose (0.1 per cent) before oral administration.

2.4.2. Eddy's Hot Plate Mediated Pain Reaction

The hot-plate test was performed to measure response latencies according to the method described by Eddy and Leimback (1953). Male Swiss albino mice were divided into four groups of six animals each. Group 1 served as control; group 2 served as standard which received morphine (10 mg/kg); groups 3 and 4 served as plant extract at a dose of 200 mg/kg and 400 mg/kg respectively. The animals were placed on the hot plate, maintained at 55±1°C. The pain threshold is considered to be reached when the animals lift and lick their paws or attempt to jump out of the hot plate. The time taken for the mice to react in this fashion was obtained using a stopwatch and noted as basal reaction time (0 min). A latency period of 15 seconds (cut-off) was defined as complete analgesia and the measurement was terminated if it exceeded the latency period in order to avoid injury (Awaad *et al.*, 2011). The reaction time was reinvestigated at 30, 60 and 120 min after the treatment and changes in the reaction time were noted.

2.5. Statistical Analysis

All the results were expressed as mean± SEM. Statistical significance was determined by using the one way ANOVA followed by Dunnett's multiple comparison tests. $p < 0.05$ was considered statistically significant.

3. Results

3.1. Acute Toxicity

In the acute toxicity studies; four mice were administered with methanolic leaves extract in graded doses of 100, 500, 1000 and 2000 mg/kg p.o., respectively. The animals were kept under observation for the change in behavior or death up to 14 days following the drug administration. The extract administration neither caused any significant change in the behaviors nor the death of animals in all the test groups. This indicates that the methanolic leaves extract of *R. ellipticus* was safe up to a single dose of 2000 mg/kg body weight. Hence we had selected 200 and 400 mg/kg oral doses to evaluate analgesic activities.

3.2. Analgesic Activity

3.2.1. Acetic Acid Induced Writhing Test

The results presented in Table 26.1 shows that the methanolic extract at the doses 200 and 400 mg/kg exhibited significant ($p<0.01$ and $p<0.001$) analgesic activity (19.4 per cent and 32.84 per cent inhibition respectively) compared to the control and also to that of aspirin, 100 mg/kg (73.13 per cent). Significant protection against writhing was observed in animals treated with aspirin, 200 and 400 mg/kg extract; where number of writhes after treatment were 18, 54 and 45 respectively compared to 67 in the control group.

Table 26.1: Effect of *R. ellipticus* leaf methanol extract on acetic acid induced writhing.

Treatment Groups	No. of Writhes (per 15 min)	Per cent Inhibition
Control	67±2.81	–
Aspirin100 (mg/kg)	18±3.05***	73.13
RELM 200 (mg/kg)	54±2.89**	19.40
RELM 400 (mg/kg)	45±3.46***	32.84

Values are expressed as mean ± SEM. (n=6), significantly different at * $p<0.05$, ** $p<0.01$, *** $p<0.001$ when compared to control. RELM - *R. ellipticus* leaf methanol

3.2.2. Eddy's Hot Plate Mediated Pain Reaction

As shown in Table 26.2, the methanolic extract produced significant analgesic activity in a dose-dependent manner. In this model, the higher dose (400 mg/kg) prolonged significantly the reaction time of animal with relatively extended duration of stimulation. At the high dose level the animals could withstand on the hot plate for 11.2, 13.6 and 7.7 seconds at 30, 60 and 120 min reaction time which was the highest and comparable with that of the reference drug morphine10 mg/kg (7.8, 9.6 and 12.4 sec.). The basal reaction time of the higher dose and standard drug were 6 and 5.8 seconds.

Table 26.2: Effect of *R. ellipticus* leaf methanol extract on hot plate mediated pain reaction.

Groups	Basal Reaction Time 15 min cut off (after drug administration)	Reaction Time 15 min cut off (after drug administration)		
		30 min	60 min	120 min
Control	7.2±0.23	9.4±0.43	8.8±0.53	9.8±0.6
Morphine (10mg/kg)	5.8±0.87	7.8±0.34***	9.6±0.8***	12.4±0.1***
RELM200 (mg/kg)	5.4±0.01	8.4±0.1***	9.4±0.1***	6.6±0.2**
RELM400 (mg/kg)	6±0.5	11.2±0.23***	13.6±0.1***	7.7±0.4**

Values are expressed as mean ± SEM. (n=6), significantly different at * $p<0.05$, ** $p<0.01$, *** $p<0.001$ when compared to control. RELM - *R. ellipticus* leaf methanol.

4. Discussion

The great antioxidant activity found previously in *R. ellipticus* as well as the traditional use encouraged us to extend our evaluation using *in vivo* models of analgesic studies. It is well known that all the pharmaceutical companies are now interested in developing more effective drugs to treat pain.

Acetic acid-induced writhing is a non-specific pain model and many compounds belonging to diverse pharmacological categories including opioids, non-steroidal anti-inflammatory drugs, calcium channel blockers, anticholinergics, antihistamines, and corticosteroids show analgesic activity in this test (Vogel and Vogel, 1997). Acetic acid test is a visceral pain model produces a painful reaction and acute inflammation in the peritoneal area. Release of arachidonic acid and biosynthesis of prostaglandin via cyclooxygenase pathway plays a role in the nociceptive mechanism of this test (Franzotti *et al.*, 2000). The analgesic effect of the tested compounds may be mediated through inhibition of cyclooxygenase and/or lipooxygenase (and other inflammatory mediators) (Vogel and Vogel, 1997; Koster, 1957). Aspirin offers relief from inflammatory pain by suppressing the formation of pain mediators in the peripheral tissues, where prostaglandins and bradykinins were suggested to play an important role in the pain process. Prostaglandins elicit pain by the direct stimulation of sensory nerve endings (Das and Ahmed, 2012). It is evident from the study that *R. ellipticus* exhibits significant peripheral analgesic effect in mice comparable with standard.

The classic hot plate model was followed to evaluate the analgesic activity of *R. ellipticus* leaf methanol extract. The hot plate model has been found suitable to investigate central antinociceptive activity because of several advantages, particularly the sensitivity to antinociceptives and limited tissue damage (Kou *et al.*, 2005). Proinflammatory mediators like prostaglandins and bradykinins were suggested to play an important role in analgesia (Vinegar *et al.*, 1969). The obtained results confirmed that leaf methanol extract at the dose 200 and 400 mg/kg has a central analgesic effect, which was compared with reference drug (Aspirin100 mg/kg). The analgesic effect of *R. ellipticus* might be attributed to the inhibition of the synthesis of some pro-inflammatory mediators, such as prostaglandins and cytokines.

5. Conclusion

In conclusion, the results of the present study revealed the strong central and peripheral analgesic activity of the leaf methanol extract of *R. ellipticus*. The data reported in this study reconfirms the traditional use of *R. ellipticus* in the treatment of various disorders. There is an urgent need for further studies to determine the mechanism behind these activities and the exploration of the exact compound or compounds responsible for the specific action.

References

Awaad, A.S., El-meligy, R.M., Qenawy, S.A., Atta, A.H. and Soliman, G.A., 2011. Anti-inflammatory, antinociceptive and antipyretic effects of some desert plants. *Journal of Saudi Chemical Society*, 15: 367–373.

Calixto, J.B., 2005. Twenty five years of research on medicinal plants in Latin America: a personal view. *Journal of Ethnopharmacology*, 100: 131–134.

Das, B.N. and Ahmed, M., 2012. Analgesic activity of the fruit Extract of *averrhoa carambola*. *International Journal of Life Sciences Biotechnology and Pharma Research*, 1(3).

Eddy, N.B. and Leimback, D., 1953. Synthetic analgesics. II. Dithyienylbutenylamines and dithyienylbutylamines. *Journal of Pharmacology and Experimental Therapeutics*, 3: 544-547.

Finn, C.E., 2008. *Rubus* spp., blackberry. In: *The Encyclopedia of Fruits and Nuts*, (Eds.) J. Janick and R.E. Paull. CABI, Cambridge, MA, pp. 348–351.

Franzotti, E.M., Santos, C.V., Rodrigues, H.M., Mourao, R.H., Andrade, M.R. and Antoniolli, A.R., 2000. Anti-inflammatory, analgesic activity and acute toxicity of *Sida cordifolia* L. (Malva-branca). *Journal of Ethnopharmacology*, 72: 273-277.

Koster, R., Anderson, M. and De Beer, E.J., 1959. Acetic acid for analgesic screening. *Federation Proceedings*, 18: 417.

Kou, J., Ni, Y., Li, N., Wang, J., Liu, L. and Jiang, Z.H., 2005. Analgesic and anti-inflammatory activities of total extract and individual fractions of Chinese medicinal plant *Polyrhachis lamellidens*. *Biological Pharmaceutical Bulletin*, 28: 176–180.

Maity, D., Pradhan, N. and Chauhan, A.S., 2004. Folk uses of some medicinal plants from north Sikkim. *Indian Journal of Traditional Knowledge*, 3: 66–71.

OECD. Guidelines for testing of chemicals, acute oral toxicity - acute toxic class method. Paris: OECD; 2001. Available from: http://iccvam.niehs.nih.gov/ SuppDocs/FedDocs/OECD/ OECD_GL423.pdf.

Patel, A.V., Rojas-Vera, J. and Dacke, C.G., 2004. Therapeutic constituents and actions of *Rubus* species. *Current Medicinal Chemistry*, 11: 1501–1512.

Pfoze, N.L., Kumar, Y. and Myrboh, B., 2012. Survey and assessment of ethnomedicinal plants used in Senapati District of Manipur State, Northeast India. *Phytopharmacology*, 2: 285–311.

Pradhan, B.K. and Badola, H.K., 2008. Ethnomedicinal plant use by Lepcha tribe of Dzongu valley, bordering Khangchendzonga Biosphere Reserve, in North Sikkim, India. *Journal of Ethnobiology and Ethnomedicine*, 4: 22 doi: 10.1186/1746-4269-4-22.

Uprety, Y., Ram, C., Asselin, P.H. and Boon, E., 2011. Plant biodiversity and ethnobotany inside the projected impact area of the Upper Seti Hydropower Project, Western Nepal. *Environ Dev Sustain.*, 13: 463–492 doi: 10.1007/s10668-010-9271-7.

Vinegar, R., Schreiber, W. and Hugo, R., 1969. Biphasic development of carrageenan edema in rats. *Journal of Pharmacology and Experimental Therapeutics*, 166: 96–103.

Vogel, H.G. and Vogel, W.H., 1997. *Drug Discovery and Evaluation*. Springer, Berlin, pp. 376–377.

Vyas, S., Agrawal, P.R., Solanki, P. and Trivedi, P., 2008. Analgesic and anti-inflammatory activities of *Trigonella foenum*-graecum (seed) extract. *Acta Poloniae Pharmaceutica*, 65: 473–476.

Scientific Basis of Herbal Medicine (2013)
Editor: Dr. Parimelazhagan Thangaraj
Published by: DAYA PUBLISHING HOUSE, NEW DELHI

Pages 235–242

Chapter 27

Anti-inflammatory Activity of Polyphenolic Extract from the Fruits of *Muntingia calabura* L. (Jamaican Cherry)

R. Gomathi[1], S. Manian[1] and N. Anusuya

[1]*Department of Botany, School of Life Sciences,
Bharathiar University, Coimbatore – 641 046, Tamil Nadu, India*
[2]*Department of Life Science, Manian Institute of Science and Technology,
Coimbatore – 641 004, Tamil Nadu, India*

1. Introduction

Polyphenols are group of structurally diverse compounds found in fruits, vegetables and plant – based food products such as tea, wine and chocolate. They have been acclaimed as natural health – protecting agents against free radical induced damages, inflammations, heart diseases, cancer and alzheimer's disease. Among several fruits, cherries are found to occupy an important position in nutritional cycle. They are found to be good sources of fiber, phenolic acids, flavonoids, anthocyanins, vitamin C, carotenoids, lutein and beta-carotene. These bioactive compounds in cherries are found to influence protective role against diabetes, inflammation, cancer and several others (McCune *et al.*, 2011).

Muntingia calabura L. (Jamaican cherry), the sole species in the genus *Muntingia*, is an evergreen tree belonging to family Elaeocarpaceae. The tree bears sweet red cheeries fruiting all over the year. Fully ripe fruits taste like cotton candy and are often cooked in tarts and made into jams. The ripe fruits are found to contain volatile phenolics, sesquiterpene and furanoid (Wong *et al.*, 1996). Furthermore, the phytocompounds such as anthocyanins, phenolic acids and flavonoids found in the fruits have demonstrated antioxidant activity (Einbond *et al.*, 2004; Kubola *et al.*, 2011). Several reports have recorded antitumor (Su *et al.*, 2003), antinociceptive, anti-inflammatory, antipyretic (Zakaria *et al.*, 2007), antimicrobial (Zakaria *et al.*, 2010), cardioprotective (Nivethetha *et al.*, 2009) and antityrosinase (Balakrishnan, 2011) activities from the leaves, roots and flowers of *M. calabura*. However, the functional property of fruits as anti-inflammatory agents has not been fully explored. The objective of the present study attempted to evaluate the anti-inflammatory activities of fruit polyphenols from *M. calabura*.

2. Materials and Methods

2.1. Plant Material

The ripe fruits of *M. calabura* were collected from trees growing in Salem, Tamil Nadu, India. Fruits were cleaned to remove adhering dust and washed in running tap water. Immediately they were freeze dried in a lyophilizer (Vir Tis Benchtop K, USA) and stored at –20°C prior to further extraction and analysis.

2.2. Chemicals

Chemicals and standard drugs were purchased from Sigma Aldrich chemical company (St. Louis, USA) or Himedia (Mumbai, India). All other reagents used were of highest purity and analytical grade made in India.

2.3. Extraction of Polyphenols

Polyphenolic contents from the fruits of *M. calabura* were extracted, according to the modified method of Sun *et al.* (2009). Twenty gram of powdered fruit sample was extracted with 100 ml of extracting solvent mixture (methanol/acetone/water, 3.5: 3.5: 3, v/v/v) containing 1 per cent formic acid for 30 min in a shaking incubator (Orbitek, 4656Z, India) at 20°C. The extraction was performed twice and the pooled extract was centrifuged at 7000 g for 15 min. The supernatant was collected and methanol and acetone were evaporated at 35°C under reduced pressure (Rotary evaporator, Yamato B0410, Japan). The lipophilic pigments in the aqueous phase were then removed with a two-fold volume of petroleum ether by two successive extractions and was further extracted thrice in ethyl acetate (ethylacetate : aqueous phase = 1:1, v/v) in a separatory funnel. The solvent was evaporated under vacuum at 35°C. The resulting residue was redissolved in ethanol and stored at -20°C for further investigation.

2.4. Phytochemical Analysis–GC-MS Analysis

The composition of phytocompounds in the polyphenolic extract of *M. calabura* were analyzed by GC-MS (Shimadzu QP2010, Europe), equipped with a Resteck-624MS column (30 m x 0.32 mm i.d., 1.8 µm film thickness). The oven temperature was programmed as follows: 45°C for 1 min, then increased by 4°C/min to 240°C and held for 2 min. The other parameters were as follows: injection temperature, 140°C; ion source temperature, 200°C; EI, 70 eV; carrier gas, He at 1.49 ml/min; injection volume, 1 µl (1/10, v/v, in ethanol) in split mode and mass range, 35-400 *m/z*. The percentage peak area obtained were quantified and identified based on computer matching with the Wiley, Nists and Fame libraries as well as with those reported in the literature.

2.5. Determination of Bioactive Compounds in the Polyphenolic Extract

Standard spectrophotometric methods were employed for the determination of non-enzymatic antioxidants in the sample extract. The antioxidant vitamins, vitamin C (Yen and Chen, 1996) and vitamin E (Prieto *et al.*, 1999) were quantified and the results were expressed as ascorbic acid (AAE) and α-tocopherol equivalents (TE), respectively. Total phenolic was calculated as gallic acid equivalent (GAE) via, Siddhuraju and Manian (2007). The method of Zhishen *et al.*, (1999) was used to obtain the total flavonoid content and the amount were expressed in terms of quercetin equivalent (QE). Total anthocyanin content was evaluated as cyanidin 3-glucoside equivalent (CGE) performed by the pH differential method of Moyer *et al.* (2002).

2.6. *In vivo* Anti-inflammatory Activity

2.6.1. Animals

Male wistar rats (120-150 g) used in this study were housed in groups of six (n = 6) per standard cage under temperature 23 ± 1°C, 12 h light/ dark cycles with food (standard pellet diet) and water *ad libitum*. Animals fasted during experiments had free access to water, but food withdrawn for 12 h before experimentation. The protocols were in accordance with the ethical guidelines for the care and use of laboratory animals, published by the CPCSEA, India (1454/PO/C/11/CPCSEA).

2.6.2. Carrageenan Induced Paw Edema

Acute anti-inflammatory activity of *M. calabura* polyphenols was determined by carrageenan induced paw edema as described by Vasudevan *et al.* (2006). The animals were divided into five groups and were fasted overnight prior to the start of the experiment with water *ad libitum*. Inflammation was induced by injecting 0.1 ml of 1 per cent carrageenan (in 1 per cent CMC, w/v) into the subplantar tissue of the right hind paw to four groups while the fifth group was maintained as untreated control. Fruit polyphenolic extracts were administered orally at doses of 200 and 400 mg/kg to the test groups of rats. As reference, induced group received distilled water (10 ml/ kg, p.o.) while the standard indomethacin (8 mg/kg, p.o.) treated animals were used as positive control. All the animals received the test sample or dosing vehicle 1 h prior to the administration of carrageenan. The thickness (mm) of the paw as a result of edematous response was measured at 0, 30, 60, 120 and 240 min intervals after carrageenan treatment using a vernier caliper (Mututoyo Digimatic Caliper, 2061 Japan) and the percentage protection was calculated.

2.6.3. Cotton Pellet–Induced Granuloma

Chronic inflammatory model using the cotton pellet implantation method was employed to study the anti-inflammatory activity of polyphenolic extract of *M. calabura* as described by Winter *et al.* (1962). The animals were divided into five groups and were fasted overnight prior to the start of the experiment with water *ad libitum*. Under ether, the rats were anaesthetized and sterilized cotton pellets (5 mg) were implanted subcutaneously through a skin incision in the back of the animals. Fruit extracts at three different doses, 100, 200 and 400 mg/kg (low, medium and high doses, respectively) and standard naproxen (25 mg/kg) were administered orally for 5 days, from the day 30 min after cotton implantation. Induced group received distilled water (10 ml/kg). On the fifth day, the granulomas were dissected out free from extraneous tissue, dried for 24 h at 60°C and the dry weights determined. Mean weight of granuloma tissue (difference between the initial and final weights) formed around each pellet and the protection percentage were calculated.

2.7. Statistical Analysis

The results were expressed as mean ± SD. The significant differences (*$P < 0.01$) in means were statistically analyzed using one-way analysis of variance (ANOVA) followed by Dunnet's t-test.

3. Results and Discussion

The chemical profile detected in GC-MS analysis, revealed the presence of twenty five constituents in the polyphenolic extract of *M. calabura* (Table 27.1). The most abundant one was phytol, an acyclic diterpene alcohol accounting for 26.26 per cent, followed by carboxylic acids such as n-hexadecanoic acid and cyclopropaneoctanoic acid (11.97 and 10.26 per cent, respectively). In addition, isoprenoids including gamma-sitosterol (11.15 per cent), stigmasterol (7.20 per cent) and campesterol (4.47 per cent) were also detected as main constituents in the fruit polyphenolic extract. In the same way, several other components were also found at appreciable contents, namely ethyl linolenate (4.48 per cent), ethyl hexadecanoate (3.65 per cent), ethyl linoleate (3.14 per cent), neophytadiene (3.03 per cent) and isoamyl acetate (2.32 per cent). From Table 27.1, it could be seen that the analyzed sample extract showed the presence of many terpenoids and carboxylic groups of chemical compounds which show significant biological activities.

Although GC-MS data did not provide information about the non-volatile phenolic groups, the presence of phenolic acids including hydrobenzoic acids (gallic acid, protocatechuic acid, p-hydroxy benzoic acid and vanillic acid) and hydrocinnamic acids (chorogenic acid, caffeic acid, syringic acid, p-coumaric acid, ferulic acid and sinapicnic acid) found in the fruits of *M. calabura* by HPLC analysis has been reported earlier (Kubola et al., 2011). Further from the results in the present investigation, the total phenolics, flavonoids and anthocyanins estimated were 121.1±0.2 mg gallic acid equivalents (GAE), 173.2±7.1 mg rutin equivalents (RE) and 82.4±1.2 mg cyanidin 3-glycoside equivalents (CGE), respectively per gram extract. The tested extract also recorded appreciable amounts of antioxidant vitamins C and E (33.6±0.4 mg ascorbic acid equivalents (AAE) and 14.7±0.7 mg α-tocopherol equivalents (TE), respectively per gram extract. The results indicated that flavonoids are the major component in the polyphenolic extract of *M. calabura*.

The anti-inflammatory activity of *M. calabura* against the carrageenan - induced edema in hind paws of rats is shown in Table 27.2. Carrageenan induced paw edema remained even 3 h after its injection into the subplantar region of paw edema in the induced group. The tested fruit polyphenol extract showed a significant (P<0.01) reduction in the paw edema volume in a dose-dependent manner. The polyphenol extract of *M. calabura* at the dose of 400 mg/kg exhibited the highest anti-inflammatory

Table 27.1: Components of *M. calabura* polyphenol extract as identified by GC-MS analysis.

Peak	Retention Time (min)	Compound	Area (per cent)
1	5.557	1,3,5-triazine-2,4,6 triimine	0.56
2	5.957	Isoamyl acetate	2.32
3	6.615	2,3-dihydro-3,5-dihydroxy-6-methyl-4H-pyran	0.78
4	6.947	Octanoic acid	0.54
5	7.786	2,3-dihydro-benzofuran	0.41
6	8.132	1,2,3-Propanetriol, monoacetate	0.40
7	8.373	n-Nonanoic acid	0.33
8	13.372	1-Desoxy-d-mannitol	1.86
9	15.242	Neophytadiene	3.03
10	15.692	(2E)-3,7,11,15-tetramethyl-2-hexadecen-1-ol	1.81
11	16.611	n-Hexadecanoic acid	11.97
12	16.847	Ethyl hexadecanoate	3.65
13	17.975	Phytol	26.26
14	18.313	Cyclopropaneoctanoic acid,2-ethylcyclopropyl methyl ester	10.26
15	18.427	Ethyl linoleate	3.14
16	18.490	Ethyl linolenate	4.48
17	18.705	Ethyl stearate	0.88
18	19.501	Octanoic acid, 2-dimethylamino ethyl ester	0.66
19	20.986	3-Cyclopentyl propionic acid	0.75
20	25.294	gamma-Tocopherol	0.82
21	25.868	alpha-Tocopherol	1.27
22	25.933	beta-Cholest-5-en-3-ol	0.97
23	26.853	Campesterol	4.47
24	27.089	Stigmasterol	7.20
25	27.711	gamma-Sitosterol	11.15

effect where it inhibited the edema formation to the extent of 68.87 per cent. Indomethacin as reference drug (8 mg/kg, p.o.), produced a significant inhibitory effect (81.92 per cent) higher than that of the investigated fruit polyphenol extracts. Carrageenan induced hind paw edema has been widely used as experimental model of acute inflammation. The development of edema induced by carrageenan is biphasic event: the early phase (1 h) involves the release of serotonin, histamine and kinins while the late phase (over 1 h) is mediated by prostaglandins, the cyclooxygenase products, and the continuity between the two phases is provided by kinins (Vinegar *et al.*, 1969). The fruit polyphenol extract of *M. calabura* exhibited a moderate inhibitory effect at early phase but was able to effectively inhibit the increase of paw volume during the late phase (3 h after carrageenan injection) of inflammation. Based on this observation, it can be suggested that this pharmacological property might be attributed to the inhibition of the release of pro-inflammatory mediators of acute inflammation especially the prostaglandins.

Table 27.2: Anti-inflammatory activity of *M. calabura* polyphenol extract on carrageenan induced paw edema in rats.

Treatment Group	Dose	Initial Paw Volume at 0 min (mm)	Final Paw Volume After 240 min (mm)	Increase in Paw Volume After 240 min (mm)	Protection (per cent)
Untreated		4.65	4.65	–	–
Induced		4.47	5.30	0.83	–
Indomethacin	8 mg/kg	4.64	4.79	0.15	81.92
MCP	200 mg/kg	4.75	5.26	0.51*	38.55
MCP	400 mg/kg	4.59	4.84	0.25*	68.87

Values are means of three independent analyses ± standard deviation (n = 6).

*$P<0.01$ significant difference compared to corresponding control.

Induced (vehicle) – distilled water; MCP – *M. calabura* polyphenol extract.

Table 27.3: Anti-inflammatory activity of *M. calabura* polyphenol extract on cotton pellet-induced granuloma in rats.

Treatment Group	Dose	Weight of Cotton Pellet (mg)	Protection (per cent)
Induced		135.39±4.2	–
Naproxen	25 mg/kg	48.36±1.7	64.28
MCP	100 mg/kg	110.72±6.2*	18.22
MCP	200 mg/kg	85.15±4.6*	37.11
MCP	400 mg/kg	60.05±3.9*	55.64

Values are means of three independent analyses ± standard deviation (n = 6).

*$P < 0.01$ significant difference compared to corresponding control. Induced (vehicle) – distilled water; MCP – *M. calabura* polyphenol extract.

The results of the chronic anti-inflammatory effect of the tested extracts of fruit polyphenols of *M. calabura* against cotton pellet induced granuloma are presented in Table 27.2. All tested extracts and reference drug showed a significant (P< 0.01) reduction in weight of cotton pellet granuloma in a dose-dependent manner (Table 27.3). The results given in Table 27.3 demonstrate that the polyphenol extract of *M. calabura* at the dose of 400 mg/kg was capable to show the maximum granuloma inhibition (55.64 per cent), which was lower to the reference drug indomethacin (64.28 per cent). The results reflected the efficacy of *M. calabura* polyphenol extract to a high extent in containing an increase in the number of fibroblasts and synthesis of collagen and mucopolysaccharide which are natural proliferative events of granulation tissue formation (Joseph *et al.*, 2010).

The flavonoids and other bioactive components in the polyphenolic extract could be attributed to its anti-inflammatory effects. Research have demonstrated the anti-inflammatory activity of flavones (apigenin, luteolin), flavonols (rutin, quercetin, kaempferol, myricetin) and many terpenoids, which can inhibit the expression of COX and lipopolysaccharides (LPS)-induced NO production (Hämäläinen *et al.*, 2007). Gabay *et al.* (2010) reported that stigmasterol was capable of counteracting inflammation and exhibit a potential anti-osteoarthritic property than campesterol and β-sitosterol, the natural anti-inflammatory phytosterols found in plants.

4. Conclusion

The data reported in this work confirmed the fruits of *M. calabura* attenuate inflammation related disorders. This can be substantiated from their significance level of polyphenols present in the fruits. Therefore, the fruits of *M. calabura* rich in antioxidant compounds might serve as a food supplement with potential pharmaceutical applications.

References

Balakrishnan, K.P., 2011. Tyrosinase inhibition and anti-oxidant properties of *Muntingia calabura* extracts: *In vitro* studies. *I. J. Pharma Biosci.*, 2: 294–303.

Einbond, L.S., Reynertson, K.A., Luo, X.D., Basile, M.J. and Kennelly, E.J., 2004. Anthocyanin antioxidants from edible fruits. *Food Chem.*, 84: 23–28.

Gabay, O., Sanchez, C., Salvat, C., Chevy, F., Breton, M., Nourissat, G., Wolf, C., Jacques, C. and Berenbaum, F., 2010. Stigmasterol: A phytosterol with potential anti-osteoarthritic properties. *Osteoarthr. Cartil.*, 18: 106–116.

Hämäläinen, M., Nieminen, R., Vuorela, P., Heinonen, M. and Moilanen, E., 2007. Anti-inflammatory effects of flavonoids: genistein, kaempferol, quercetin, and daidzein inhibit STAT-1 and NF-kappaB activations, whereas flavone, isorhamnetin, naringenin, and pelargonidin inhibit only NF-kappaB activation along with their inhibitory effect on iNOS expression and NO production in activated macrophages. *Mediat. Inflamm.*, 2007: 1–10.

Joseph, J.M., Sowndhararajan, K. and Manian, S., 2010. Evaluation of analgesic and anti-inflammatory potential of *Hedyotis puberula* (G.Don) R. Br. ex Arn. in experimental animal models. *Food Chem. Toxicol.*, 48: 1876–1880.

Kubola, J., Siriamornpun, S. and Meeso, N., 2011. Phytochemicals, vitamin C and sugar content of Thai wild fruits. *Food Chem.*, 126: 972–981.

McCune, L.M., Kubota, C., Stendell-Hollis, N.R. and Thomson, C.A., 2011. Cherries and health: A review. *Crit. Rev. Food Sci. Nutr.*, 51: 1–12.

Moyer, R.A., Hummer, K.E., Finn, C.E., Frei, B. and Wrolstad, R.E., 2002. Anthocyanins, phenolics, and antioxidant capacity in diverse small fruits: Vaccinium, rubus, and ribes. *J. Agric. Food Chem.*, 50: 519–525.

Nivethetha, M., Jayasri, J. and Brindha, P., 2009. Effects of *Muntingia calabura* L. on isoproterenol-induced myocardial infarction. *Singapore Med. J.*, 50: 300–302.

Prieto, P., Pineda, M. and Aguilar, M., 1999. Spectrophotometric quantitation of antioxidant capacity through the formation of a phosphomolybdenum complex: specific application to the determination of vitamin E. *Anal. Biochem.*, 269: 337–341.

Siddhuraju, P. and Manian, S., 2007. The antioxidant and free radical scavenging capacity of dietary phenolic extracts from horse gram (*Macrotyloma uniflorum* (Lam.) Verdc.) seeds. *Food Chem.*, 105: 950–958.

Su, B.N., Park, E.J., Vigo, J.S., Graham, J.G., Cabieses, F., Fong, H.H.S., Pezzuto, J.M. and Kinghorn, A.D., 2003. Activity-guided isolation of the chemical constituents of *Muntingia calabura* using a quinine reductase induction assay. *Phytochem.*, 63: 335–341.

Sun, J., Yao, J., Huang, S., Long, X., Wang, J., Garcßa, E.G., 2009. Antioxidant activity of polyphenols and anthocyanin extracts from fruits of *Kadsura coccinea* (Lem.) A.C. Smith. *Food Chem.*, 117: 276–281.

Vasudevan, M., Gunnam, K.K. and Parle, M., 2006. Antinociceptive and anti–inflammatory properties of *Daucus carota* seeds extract. *J. Health Sci.*, 52: 598–606.

Vinegar, R., Schreiber, W. and Hugo, R., 1969. Biphasic development of carrageenan edema in rats. *J. Pharmacol. Exp. Ther.* 166: 96–103.

Winter, C.A., Risley, E.A. and Nuss, G.W., 1962. Carrageenan induced edema in hind paw of rats as an assay for anti-inflammatory drugs. *Proc. Soc. Exp. Biol. Med.*, 11: 544–547.

Wong, K.C., Chee, S.G. and Er, C.C., 1996. Volatile constituents of the fruits of *Muntingia calabura*. *J. Essent. Oil Res.*, 8: 423–426.

Yen, G.C. and Chen, H.Y., 1995. Antioxidant activity of various tea extracts in relation their antimutagenicity. *J. Agri. Food Chem.*, 43: 27–32.

Zakaria, Z.A., Hazalin, N.A.M.N., Zaid, S.N.H.M., Ghani, M.A., Hassan, M.H., Gopalan, H.K. and Sulaiman, M.R., 2007. Antinociceptive, anti-inflammatory and antipyretic effects of *Muntingia calabura* aqueous extract in animal models. *J. Nat. Med.*, 61: 443–448.

Zakaria, Z.A., Mohamed, A.M., Mohd Jamil, N.S., Rofiee, M.S., Hussain, M.K., Sulaiman, M.R., Teh, L.K. and Salleh, M.Z., 2010. *In vitro* antiproliferative and antioxidant activities of the extracts of *Muntingia calabura* leaves. *Am. J. Chin. Med.*, 39: 183–200.

Zhishen, J., Mengcheng, T. and Jianming, W., 1999. The determination of flavonoid contents in mulberry and their scavenging effects on superoxide radicals. *Food Chem.*, 64: 555–559.

Scientific Basis of Herbal Medicine (2013)
Editor: Dr. Parimelazhagan Thangaraj
Published by: DAYA PUBLISHING HOUSE, NEW DELHI

Pages 243–250

Chapter 28

Total Phenolic Content and Anti-Radical Activity of Mixed Extracts of *Caralluma* spp.

Rahul Chandran and Thangaraj Parimelazhagan

Bioprospecting laboratory, Department of Botany,
Bharathiar University, Coimbatore, Tamil Nadu, India

1. Introduction

Herbal medicine is now globally accepted as a valid alternative system of therapy in the form of pharmaceuticals, functional foods, etc. To trace the history of phytotherapy is to trace the history of humanity itself. The discovery of the curative properties of certain plants must have sprung from instinct. As a multidisciplinary science the research in this field is almost unlimited, which makes it impractical to discuss all the aspects of this emerging science in just a sentence. Free radicals, *e.g.*, various reactive oxygen species (ROS), are ubiquitously produced during cellular metabolism in living systems and they are responsible for a number of oxidative stress related disorders in human beings, such as atherosclerosis, ageing, cancer and cardiovascular diseases (Moein *et al.*, 2012). Both edible and non-edible parts of plants constitute phenolic compounds. These compounds delay the oxidation of various compounds "important for life" by inhibiting the initiation or propagation of oxidising chain reactions (Amarowicz *et al.*, 2010). Phenolic acids have attracted increasing attention for their antioxidant behaviour and beneficial health-promoting effects and they account for about one-third of the phenolic compounds in plant foods. It is assumed that many antioxidative phenolic compounds in plants are usually presented in a covalently-bound form (Xu *et al.*, 2007). These natural antioxidants not only protect food lipids from oxidation, but also provide health benefits associated with preventing damage due to biological degeneration (Hu and Dendelion, 2005).

Caralluma diffusa and *Caralluma adscendens*, family Asclepiadaceae is consumed by the tribes and believed to reduce obesity (Ramachandran, *et al.*, 2011). *Caralluma* species have been used as emergency food for centuries in semi-arid areas of Pakistan

(Bnouham *et al.*, 2003). Apart from this, *Caralluma* species have shown anti-inflammatory (Zakaria *et al.*, 2001), gastric mucosa protecting, anti-ulcer properties (Al-Harbi *et al.*, 1994), antihyperglycemic activity (Jayakar, *et al.*, 2003). In Indian system of tradition medicine there has been a practice of polyherbal formulations to cure several diseases. Hence a similar attempt has been made to mix the whole plant of *C. diffusa* and *C. adscendens* to extract biologically active compounds and to find their combined effect. Hence the present study was done to estimate the total phenolic and antioxidant properties to support its value in treatment against free radical generated diseases.

2. Materials and Methods

2.1. Collection of Plant Materials

Carralluma diffusa and *Caralluma adscendens* were collected from Madukkarai, Coimbatore district during the month of May-June 2011. The collected plant material was identified and their authenticity was confirmed by comparing the voucher specimen at the herbarium of Botanical survey of India, Southern circle, Coimbatore, Tamil Nadu. Freshly collected plant material was cleaned to remove adhering dust and then dried under shade. The dried sample were powdered and used for further studies.

2.2. Chemicals

Potassium ferricyanide, ferric chloride, 2,2-diphenyl-1-picryl-hydrazyl (DPPH), potassium persulfate, 2,2'aninobis(3-ethylbenzothiozoline-6-sulfonic acid) disodium salt (ABTS), 6-hydroxy-2,5,7,8-tetramethylchroman-2-carboxylic acid (Trolox), ferrous chloride, 2,4,6-tripyridyl-s-triazine (TPTZ), polyvinyl polypyrrolidone (PVPP), ethylenediamine tetracetic acid (EDTA) disodium salt, 2,2'-bipyridyl and hydroxylamine hydrochloride, NBT, BHT. All other reagents used were of analytical grade.

2.3. Successive Solvent Extraction

The air dried, powdered whole part of *C. diffusa* and *C. adscendens* were mixed thoroughly through quadrant mixing method and was extracted in soxhlet extractor successively with petroleum ether, ethyl acetate and methanol. Each time before extracting with the next solvent, the material was dried in hot air oven below 40°C. Finally, the material was macerated using hot water with occasional stirring for 24 hr and the water extract was filtered. The different solvent extracts were concentrated by rotary vacuum evaporator and then air dried. The dried extract obtained with each solvent was weighed. The percentage yield was expressed in terms of air dried weight of plant material. The extracts thus obtained were used directly for the estimation of total phenolics and also for the assessment of antioxidant potential through various biochemical assays. The extracts were freeze dried and stored in desiccators until further analysis.

2.4. Determination of Total Phenolics and Tannins

The total phenol content was determined according to the method described by Siddhuraju and Becker, 2003. Triplicate concentration of leaf extract (2 mg/2 mL) was taken in the test tubes and made up to the volume of 1 mL with distilled water. Then 0.5mL of Folin - Ciocalteu reagent (1:1 with water) and 2.5 mL of sodium carbonate solution (20 per cent) were added sequentially in each tube. Soon after vortexing the reaction mixture, the test tubes were placed in dark for 40 min and the absorbance was recorded at 725 nm against blank. Reaction mixture without plant extract was taken as blank. The analysis was performed in triplicate and the results were expressed in gallic acid equivalents.

Using the same extract the tannins were estimated after treatment with polyvinyl polypyrrolidine (PVPP) (Sidduraju and Manian, 2007). 100 mg of PVPP was weighed into a 100 × 12 mm test tube and to this 500 µL distilled water and then 500 µL of the sample extract was added. The content was vortexed and kept in the test tube at 4°C for 15 min. Then the sample was centrifuged at 4000 rpm for 10 min at room temperature and the supernatant was collected. This supernatant has only simple phenolics other than the tannins (the tannins would have been precipitated along with the PVPP). The phenolic content of the supernatant was measured and expressed as the content of non-tannin phenolics on a dry matter basis. From the above results, the tannin content of the sample was calculated as follows:

Tannin (per cent) = Total phenolics (per cent) - Non tannin phenolics (per cent)

2.5. *In vitro* Antioxidant Studies

2.5.1. Total Antioxidant Activity by 2, 2'-azinobis (3-Ethylebenzothiozoline-6-sulphonic acid) (ABTS•+) Radical Cation Assay

The total antioxidant activity of the extract was measured by ABTS radical cation decolorization assay according to the method of Re *et al.,* 1999 described by Siddhuraju and Manian, 2007. ABTS•+ was produced by reacting 7 mM ABTS•+ aqueous solution with 2.4mM potassium persulphate in the dark for 12 – 16 hr at room temperature. Prior to assay, this solution was diluted in ethanol (about 1:89 v/v) and equilibrated at 30°C to give an absorbance at 734 nm of 0.7 ± 0.02. The stock solution of the sample extracts were diluted such that after addition of different aliquots into the assay, they produced between 20- 80 per cent inhibition of the blank (ethanol) absorbance. After the addition of 1mL of diluted ABTS•+ solution to different concentration of sample or trolox standards (final concentration 0-15 µM) in ethanol, absorbance was measured at 30°C exactly 30 min after initial mixing. Appropriate solvent blanks were also run in each assay. Triplicate determinations were made at each dilution of the standard, and the percentage inhibition was calculated of the blank absorbance at 734 nm and it was plotted as a function of trolox concentration. The unit of total antioxidant activity (TAA) is defined as the concentration of trolox having equivalent antioxidant activity expressed as µMol/g sample extract on dry matter.

2.5.2. Ferric Reducing Antioxidant Power (FRAP) Assay

The antioxidant capacities of phenolic extracts of samples were estimated according to the procedure described by Pulido *et al.,* 2000. FRAP reagent (900 µl), prepared freshly and incubated at 37°C, was mixed with 90 µl of distilled water and

50µL of test sample or distilled water (for the reagent blank). The test sample and reagent blank were incubated at 37°C for 30 min in a water bath. The final dilution of the test sample in the reaction mixture was 1/34. The FRAP reagent contained 2.5mL of 20 mmol/L TPTZ (2,4,6-tripyridyl-s-triazine) solution in 40 mmol/l HCl plus 2.5mL of 20 mmol/L $FeCl_3 \cdot 6H_2O$ and 25 mL of 0.3 mol/L acetate buffer (pH 3.6) described by Siddhuraju and Becker. At the end of incubation, the absorbance readings were taken immediately at 593 nm. The FRAP value is expressed as mmol Fe (II) equivalent/mg extract.

2.5.3. Metal Chelating Activity

The chelating of ferrous ions by *C. diffusa* and *C. adscendens* mixed extract was estimated by the method of Dinnis *et al.*, 1994. Briefly, 50 µl of 2 mM $FeCl_2$ was added to 1mL of triplicate concentration of the extract. The reaction was initiated by the addition of 0.2mL of 5mM ferrozine solution. The mixture was vigorously shaken and left to stand at room temperature for 10 min. The absorbance of the solution was thereafter measured at 562 nm against deionized water which was used as blank. BHT was taken as standard. All the reagents without addition of sample extract was used as negative control. The percentage inhibition of ferrozine – Fe^{2+} complex formation was calculated as follows

2.6. Statistical Analysis

All experiments were repeated at least three times. Results were reported as Mean ± SD. The statistical significance between antioxidant activity values of the extracts was evaluated with one way ANOVA followed by Holm-Sidak test.

3. Results

3.1. Total Phenolics, Tannins and Flavonoids

In Table 28.1, methanol extract shows highest phenolic content (10.423 g GAE/100 g) when compared to other extracts. The least phenolic was obtained from petroleum ether extract (2.156 g of GAE/100 g). Higher tannin content was found in ethyl acetate extract (4.46 g of GAE/100 g) when compared to other extracts with least amount in petroleum ether extract (0.51 g of GAE/100 g). The phenolics and tannin content found in these results may contribute to the antioxidant/free radical scavenging property of this mixed extracts.

Table 28.1: Phenolic and tannin in mixed extracts of C. diffusa and Caralluma adscendens.

Mixed Extract	Total Phenolic g GAE equi/100 g Extract	Tannin g GAE equi/100 g Extract
Pet. Ether	2.156±0.050	2.65±0.19
Ethyl acetate	9.087±0.198	4.46±0.22
Methanol	10.423±0.405	2.15±0.40
Hot water	4.815±0.282[c]	0.51±0.25

Data expressed as mean ± standard deviation. GAE: Gallic acid equivalents; RE: Rutin equivalents.

3.2. Total Antioxidant Activity by 2, 2'-azinobis (3-Ethylebenzothiozoline-6-Sulphonic acid) (ABTS•+) Radical Cation Assay

The results of the total antioxidant activity of different extracts are given in the Table 28.2. Hot water extract showed significant activity (703.35 µ moles TE/g extract) while the methanol extract revealed lowest (272.02 µmoles/g) ability to quench radicals. ABTS radical scavenging activity was expressed as trolox equivalent. The ABTS+ stabilizing ability of the extracts are in the order hot water > petroleum ether >ethyl acetate>methanol.

3.3. Ferric Reducing Antioxidant Power (FRAP) Assay

The results presented in Table 28.2, shows that the ferric reducing capacity of methanol extract is higher (721.1 mM Fe II/mg extract) when compared to other extracts. The FRAP assay measures the antioxidant effect of any substances in the reaction medium as reducing ability. Here the petroleum ether extract showed lowest reduction ability (376.2 mM Fe II/mg extract) compared to other extracts.

Table 28.2: ABTS+ and FRAP activity of mixed extracts of *C. diffusa* and *Caralluma adscendens*.

Mixed extract	ABTS•+ µM trolox equi/g extract	FRAP mM Fe II/mg extract
Pet. Ether	506.25±25.29	376.2±13.00
Ethyl acetate	436.38±18.75	599.8±56.95
Methanol	294.97±42.15	721.1±57.60
Hot water	703.35±5.85	678.7±48.25

Data expressed as mean ± standard deviation.

3.4. Metal Chelating Activity

The Fe^{2+} chelating activity of extracts are shown in Figure 28.1. A good chelation was observed in the methanol extract (65.6 per cent), followed by hot water extract (46.86 per cent). The results are comparably low to that of standard BHT (92 per cent). The least activity of extracts may be because of the interference of other metal ions in the extract or reaction mixture.

4. Discussion

Polyphenolic contents of all the sample extracts appear to function as good electron and hydrogen atom donor and therefore should be able to terminate radical chain reaction by converting free radical and reactive oxygen species to more stable products. It is important to determine the positive correlation between antioxidant potential and Phenolic content estimated by Folin - Ciocalteu method (Ksouri *et al.*, 2007). The amount of phenolics present in the methanol extract might prove the antioxidant potential of the plant. Hence presence of total phenols and tannins supports that the extracts can have good antioxidant property. The assay based on the use of ABTS+ radical are among the most popular spectrophotometric methods

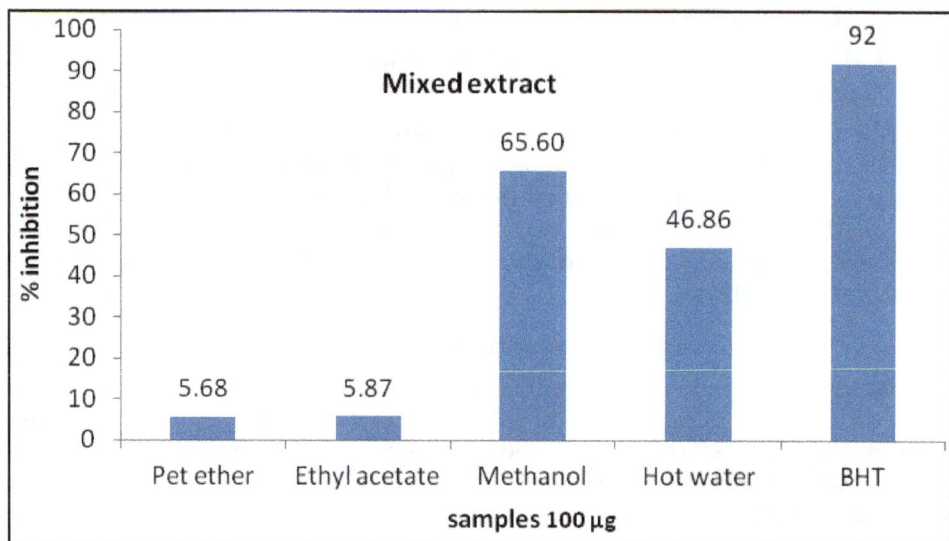

Figure 28.1: Metal chelating activity of mixed extracts of *C. diffusa* and *Caralluma adscendens*.

for the determination of the antioxidant capacity of food, beverages and vegetable extracts (Bendini *et al.*, 2006). Though the total phenolic content of the samples of respective solvent extracts found to be relatively low, metal chelating ability by the samples seems to be efficient for functioning as potential nutraceuticals. Metal ions like iron are an essential element which is necessary for transport of oxygen molecule through blood. Many studies have revealed that the plant extracts rich in phenolic content are capable of forming complexes and stabilizing transition metal ions, rendering them to unable to participate in metal catalysed initiation and hydroperoxide decomposition reactions (Bourgou *et al.*, 2008). Metal chelating ability was significant as they reduce the concentration of catalyzing transition metal in lipid peroxidation (Duh *et al.*, 1999). Commendable result shown by the methanolic extract can be taken as an efficient hydrogen donor to make the free radical a stable one. Hot water extract also showed comparatively good activity. A similar antioxidant potential was shown alone by *C. adscendens* in DPPH, lipid peroxidation and reducing power assay (Tatiya *et al.*, 2010).

5. Conclusion

The *in vitro* total phenolic content and antioxidant activity was assessed in mixed extracts of *Caralluma diffusa* and *Caralluma adscendens*. The results of this study confirm the significance as dietary sources of natural antioxidants. A good correlation between phenolic content and antioxidant activity was observed. Methanol was more efficient than other extracts in extracting antioxidant compounds from extracts. The methanol extract also showed good antioxidant capacity in almost all the antioxidant assays. Based on this work it can be concluded that this plant can be a source of antioxidant food. Studies focusing the exploration of medicinal wealth of plants have attained considerable interest in recent years. Food plants with high antioxidant property are being given special interest to fight against several degenerative diseases. These plant which is consumed by the locals and hence investigated for its total phenolic and antioxidant property for the first time.

Acknowledgement

The authors are thankful to Department of Science and Technology, Govt. of India and INSPIRE programme for providing financial support for carrying out the work.

References

Al-Harbi, M.D., Qureshi, S., Ahmed, M.M., Afzal, M. and Shah, A.H., 1994. Evaluation of *Caralluma tuberculata* pretreatment for the protection of rat gastric mucosa against toxic damage. *Toxicol. Appl. Pharmacol.,* 28: 1–8.

Amarowicz, R., Estrella, I., Hernandez, T., Robredo, S., Troszyn´ ska a A, and Kosin´ ska a A, *et al.,* 2010. Free radical-scavenging capacity, antioxidant activity, and phenolic composition of green lentil (Lens culinaris). *Food Chem.,* 121: 705–711.

Bendini, A., Cerretani, L., Pizzolante, L., Toschi, T.G., Guzzo, F. and Ceoldo, S., *et al.,* 2006. Phenol content related to antioxidant and antimicrobial activities of *Passiflora* spp. extracts. *Eur. Food Res. Tech.,* 223: 102.

Bnouham, M., Merhfour, F.Z., Ziyat, A., Mekhfi, H., Aziz, M. and Legssyer, A., 2003. Antihyperglycemic activity of the aqueous extract of *Urtica diocia. Fitoterapia,* 74: 677–681.

Bourgou, S., Ksouri, R., Bellila, A., Skandrani, I., Falleh, H. and Marzouk, B., 2008. Phenolic composition and biological activities of Tunisian *Nigella sativa* L. shoots and roots. *Compte Rendu de Biologies,* 331: 48–55.

Dinis, T.C.P., Madeira, V.M.C. and Almeida, L.M., 1994. Action of phenolic derivatives (acetoaminophen, salycilate and 5-aminosalycilate) as inhibitors of membrane lipid peroxidation and as peroxyl radical scavengers. *Arch. Biochem. Biophys.,* 315: 161–169.

Duh, P.D., Tu, Y.Y. and Yen, G.C., 1999. Antioxidant activity of water extract of harng Jyur (*Chrysanthemum morifolium* Ramat). *Lebensmittel-Wissenschaft Und technologie,* 32: 296–277.

Hu, C. and Kitts, D.D., 2005. Dendelion (*Taraxacum*) flower extract suppresses both reactive oxygen species and nitric oxide and prevents lipid oxidation *in vitro. Phytomed,* 12: 588–597.

Jayakar, B., Rajkapoor, B. and Suresh, B., 2004. Effect of *Caralluma attenuate* in normal and alloxan induced diabetic rats. *J. Herb Pharmacother.,* 4: 35–40.

Ksouri, R., Megdiche, W., Debez, A., Falleh, H., Grignon, C. and Abdelly, C., 2007. Salinity effects on polyphenol content and antioxidant activities in leaves of the halophyte Cakile maritima. *Plant Physiol. Biochem.,* 45: 244–249.

Moein, M.R., Moein, S. and Ahmadizadeh, S., 2008. Radical scavenging and reducing power of Salvia mirzayanii subfractions. *Molecules*, 13: 2804–2813.

Pulido, R., Bravo, L. and Sauro-Calixto, F., 2000. Antioxidant activity of dietary polyphenols as determined by a modified ferric reducing/antioxidant power assay. *J. Agric. Food Chem.*, 48: 3396–3402.

Ramachandran, V.S., Thomas, B., Sofiya, C. and Sasi, R., 2011. Rediscovery of an endemic species, *Caralluma diffusa* (Wight) N.E. Br. (Asclepiadaceae) from Coimbatore District, Tamil Nadu, India, after 160 years. *Journal of Threatened Taxa*, 3(3): 1622–1623.

Re, R., Pellegrini, N., Proteggente, A., Pannala, A., Yang, M. and Rice-Evans, C., 1999. Antioxidant activity applying an improved ABTS radical cation decolorization assay. *Free Radic. Biol. Med.*, 26: 1231–1237.

Siddhuraju, P. and Becker, K., 2003. Studies on antioxidant activities of Mucuna seed (*Mucuna pruriens* var. utilis) extracts and certain non-protein amino/imino acids through in vitro models. *J. Sci. Food Agric.*, 83: 1517–1524.

Siddhuraju, R. and Manian, S., 2007. The antioxidant activity and free radical scavenging capacity of dietary phenolic extracts from horse gram (*Macrotyloma uniflorum* (Lam.) Verdc.) seeds. *Food Chem.*, 105: 950–958.

Tatiya, A.U., Kulkarni, A.S., Surana, S.J. and Bari, N.D., 2010. Antioxidant and hypolipidemic effect of Caralluma adscendens Roxb. in alloxanized diabetic rats. *Int. J. Pharmacol.*, 6 (4): 400–406.

Xu, G., Ye, X., Chen, J. and Liu, D., 2007. Effect of heat treatment on the Phenolic compounds and antioxidant capacity of citrus peel extract. *J. Agri. Food Chem.*, 55: 330–335.

Zakaria, M.N., Islam, M.W., Radhakrishnan, R., Chen, H.B., Kamil, M. and Al, Gifri, et al., 2001. Antinociceptive and anti-inflammatory properties of *Caralluma arabi*. *J. Ethnopharmacol.*, 76: 155–158.

Index

* 9 7 8 9 3 5 1 3 0 0 7 4 8 *